Probability
an introduction

Geoffrey Grimmett
Statistical Laboratory, University of Cambridge

Dominic Welsh
Merton College, Oxford

CLARENDON PRESS · OXFORD

OXFORD
UNIVERSITY PRESS

Great Clarendon Street, Oxford OX2 6DP

Oxford University Press is a department of the University of Oxford.
It furthers the University's objective of excellence in research, scholarship,
and education by publishing worldwide in

Oxford New York

Auckland Cape Town Dar es Salaam Hong Kong Karachi
Kuala Lumpur Madrid Melbourne Mexico City Nairobi
New Delhi Shanghai Taipei Toronto
With offices in
Argentina Austria Brazil Chile Czech Republic France Greece
Guatemala Hungary Italy Japan South Korea Poland Portugal
Singapore Switzerland Thailand Turkey Ukraine Vietnam

Oxford is a registered trade mark of Oxford University Press
in the UK and in certain other countries

Published in the United States
by Oxford University Press Inc., New York

ISBN 978-0-19-853264-4

Printed and bound in Great Britain by
CPI Antony Rowe, Chippenham and Eastbourne

Preface

Probability and statistics are now taught widely in schools and are integral parts of many O-level and A-level syllabuses. Consequently the attitudes of universities towards these subjects have changed over the last few years and, at many universities, first-year mathematics students learn material which was previously taught in the third year only. This text is based upon first and second year courses in probability theory which are given at the Universities of Bristol and Oxford.

Broadly speaking we cover the usual material, but we hope that our account will have certain special attractions for the reader and we shall say what these may be in a moment. The first eight chapters form a course in basic probability, being an account of events, random variables, and distributions—we treat discrete and continuous random variables separately—together with simple versions of the law of large numbers and the central limit theorem. There is an account of moment generating functions and their applications. The last three chapters are about branching processes, random walks, and continuous-time random processes such as the Poisson process; we hope that these chapters are adequate at this level and are suitable appetizers for courses in applied probability and random processes. We have deliberately omitted various topics which are crucial in more advanced treatments, such as the theory of Markov chains, and we hope that most critics will agree with such decisions. In the case of Markov chains, we could not justify to ourselves the space required to teach more than mere fragments of the theory. On the other hand we have included a brief treatment of characteristic functions in two optional sections for the more advanced reader.

We have divided the text into three sections: (A) Probability, (B) Further Probability, and (C) Random Processes. In doing so we hope to indicate two things. First, the probability in Part A seems to us to be core material for first-year students, whereas the material in Part B is somewhat more difficult. Secondly, although random processes are collected together in the final three chapters, they may well be introduced much earlier in the course. The chapters on branching processes and random walks might well come after Chapter 5, and the chapter on continuous-time processes after Chapter 6.

We have two major aims: to be concise and to be honest about mathematical rigour. Some will say that this book reads like a set of lecture notes. We would not regard this as entirely unfair; indeed a principal reason for writing it was that we believe that most students benefit more from possessing a compact account of the subject in 200 printed pages or so (at a suitable price) than a diffuse account of 400 pages. Most undergraduates learn probability theory by attending lectures, at which they normally take copious and occasionally incorrect notes; they may also attend tutorials and classes. Few are the undergraduates who learn probability in private by relying on a textbook as the sole, or even principal, source of inspiration and learning. Although some will say that this book is too difficult, it is the case that first-year students at many universities learn some quite difficult things, such as axiomatic systems in algebra and ε/δ analysis, and we doubt if much of the material covered here is inherently more challenging than these. Also, lecturers and tutors have certain advantages over authors—they have the power to hear and speak to their audiences—and these advantages should help them to explain the harder things to their students.

Here are a few words about our approach to rigour. It is clearly impossible to prove everything with complete rigour at this level; on the other hand it is important that students should understand why rigour is necessary. We try to be rigorous where possible, and elsewhere we go to some lengths to point out how and where we skate over thin ice. This can occasionally be tedious.

Most sections finish with a few exercises; these are usually completely routine, and students should do them as a matter of course. Each chapter finishes with a collection of problems; these are often much harder than the exercises, and include many parts of questions taken from examination papers set in Bristol and Oxford; we acknowledge permission from Bristol University and from Oxford University Press in this regard. There is a final chapter containing some hints for solving the problems. Problems marked with an asterisk may be rather difficult.

We hope that the remaining mistakes and misprints are not held against us too much, and that they do not pose overmuch of a hazard to the reader. Only with the kind help of our students have we reduced them to the present level. Finally we thank Rhoda Rees for typing the manuscript with such skill, speed and good cheer.

Bristol and Oxford   G. G.
July 1985   D. W.

Contents

A. Basic Probability

1
Events and probabilities

1.1 Experiments with chance

Many actions have outcomes which are largely unpredictable in advance—tossing a coin and throwing a dart are simple examples. Probability theory is about such actions and their consequences. The mathematical theory starts with the idea of an *experiment* (or *trial*), being a course of action whose consequence is not predetermined; this experiment is reformulated as a mathematical object called a *probability space*. In broad terms the probability space corresponding to a given experiment comprises three items:

  (i) the set of all possible outcomes of the experiment,
 (ii) a list of all the events which may possibly occur as consequences of the experiment,
(iii) an assessment of the likelihoods of these events.

For example, if the experiment is the throwing of a fair six-sided die, then the probability space contains

  (i) the set $\{1, 2, 3, 4, 5, 6\}$ of possible outcomes,
 (ii) a list of events such as 'the result is 3',
                                'the result is at least 4',
                                'the result is a prime number',
(iii) the assessment that each number $1, 2, 3, 4, 5, 6$ is equally likely to be the result of the throw.

Given any experiment involving chance, there is a corresponding probability space, and the study of such spaces is called *probability theory*. Next, we shall see how to construct such spaces more explicitly.

1.2 Outcomes and events

We use the letter $\mathscr{E}$ to denote a particular experiment whose outcome is not completely predetermined. The first thing which we do is to make a list of all the possible outcomes of $\mathscr{E}$; the set of all such possible outcomes is called the *sample space* of $\mathscr{E}$ and we usually denote it by Ω. The Greek letter ω denotes a typical member of Ω, and we call each member ω of Ω an *elementary event*.

If, for example, $\mathscr{E}$ is the experiment of throwing a fair die once, then $\Omega = \{1, 2, 3, 4, 5, 6\}$. There are many questions which we may wish to ask about the actual outcome of this experiment (questions such as 'is the outcome a prime number?'), and all such questions may be rewritten in terms of subsets of Ω (the previous question becomes 'does the outcome lie in the subset $\{2, 3, 5\}$ of Ω?'). The second thing which we do is to make a list of all the events which are interesting to us; this list takes the form of a collection of subsets of Ω, each such subset A representing the event 'the outcome of $\mathscr{E}$ lies in A'. Thus we ask 'which possible events are interesting to us' and then we make a list of the corresponding subsets of Ω. This relationship between *events* and *subsets* is very natural, especially because two or more events combine with each other in just the same way as the corresponding subsets combine; for example, if A and B are subsets of Ω then

the set $A \cup B$ corresponds to the event 'either A or B occurs',

the set $A \cap B$ corresponds to the event 'both A and B occur',

the set $\Omega \backslash A$ corresponds† to the event 'A does not occur',

where we say that a subset C of Ω 'occurs' whenever the outcome of $\mathscr{E}$ lies in C. Thus all set-theoretic statements and combinations may be interpreted in terms of events; for example, the formula

$$\Omega \backslash (A \cap B) = (\Omega \backslash A) \cup (\Omega \backslash B)$$

may be read as 'if A and B do not both occur, then either A does not occur or B does not occur'. In a similar way, if $A_1, A_2, \ldots$ are events then the sets $\bigcup_{i=1}^{\infty} A_i$ and $\bigcap_{i=1}^{\infty} A_i$ represent the events 'A_i occurs, for some i' and 'A_i occurs, for every i', respectively.

Thus we write down a collection $\mathscr{F} = \{A_i : i \in I\}$ of subsets of Ω which are interesting to us; each $A \in \mathscr{F}$ is called an *event*. In simple cases, such as the die-throwing example above, we usually take $\mathscr{F}$ to be the set of *all* subsets of Ω (called the *power set* of Ω), but for reasons which may be appreciated later there are many circumstances in which we take $\mathscr{F}$ to be a very much smaller collection than the entire power set. In all cases we demand a certain consistency of $\mathscr{F}$, in the following sense. If $A, B, C, \ldots \in \mathscr{F}$ then we may reasonably be interested also in the events 'A does *not* occur' and 'at least one of $A, B, C, \ldots$ occurs'. With this in mind we require that $\mathscr{F}$ satisfy the following definition.

The collection $\mathscr{F}$ of subsets of the sample space Ω is called an *event space* if

(1) $\mathscr{F}$ is non-empty,

---

† For any subset A of Ω, the *complement* of A is the set of all members of Ω which are not members of A. We denote the complement of A by either $\Omega \backslash A$ or A^c, depending on the context.

(2)   if $A \in \mathcal{F}$ then $\Omega \backslash A \in \mathcal{F}$,

(3)   if $A_1, A_2, \ldots \in \mathcal{F}$ then $\bigcup_{i=1}^{\infty} A_i \in \mathcal{F}$.

We speak of an event space $\mathcal{F}$ as being 'closed under the operations of taking complements and countable unions'. An elementary consequence of axioms (1)–(3) is that an event space $\mathcal{F}$ must contain the empty set $\varnothing$ and the whole set Ω. This holds since $\mathcal{F}$ contains some set A (from (1)), and hence $\mathcal{F}$ contains $\Omega \backslash A$ (from (2)), giving also that $\mathcal{F}$ contains the union $\Omega = A \cup (\Omega \backslash A)$ together with the complement $\Omega \backslash \Omega = \varnothing$ of this last set.

Here are some examples of pairs $(\Omega, \mathcal{F})$ of sample spaces and event spaces.

Example 4 Ω is any set and $\mathcal{F}$ is the power set of Ω.   □

Example 5 Ω is any set and $\mathcal{F} = \{\varnothing, A, \Omega \backslash A, \Omega\}$ where A is a given subset of Ω.   □

Example 6 $\Omega = \{1, 2, 3, 4, 5, 6\}$ and $\mathcal{F}$ is the collection

$$\varnothing, \{1, 2\}, \{3, 4\}, \{5, 6\}, \{1, 2, 3, 4\}, \{3, 4, 5, 6\}, \{1, 2, 5, 6\}, \Omega$$

of subsets of Ω. This event space is unlikely to arise naturally in practice.   □

Exercises   In these exercises, Ω is a set and $\mathcal{F}$ is an event space of subsets of Ω.
1. If $A, B \in \mathcal{F}$, show that $A \cap B \in \mathcal{F}$.
2. The *difference* $A \backslash B$ of two subsets A and B of Ω is the set $A \cap (\Omega \backslash B)$ of all points of Ω which are in A but not in B. Show that if $A, B \in \mathcal{F}$, then $A \backslash B \in \mathcal{F}$.
3. The *symmetric difference* $A \triangle B$ of two subsets A and B of Ω is defined to be the set of points of Ω which are in either A or B but not in both. If $A, B \in \mathcal{F}$, show that $A \triangle B \in \mathcal{F}$.
4. If $A_1, A_2, \ldots, A_m \in \mathcal{F}$ and k is a positive integer, show that the set of points in Ω which belong to exactly k of the A's belongs to $\mathcal{F}$ (the previous exercise is the case when $m = 2$ and $k = 1$).
5. Show that if Ω is a finite set then $\mathcal{F}$ contains an even number of subsets of Ω.

1.3 Probabilities

From our experiment $\mathcal{E}$, we have so far constructed a sample space Ω and an event space $\mathcal{F}$ associated with $\mathcal{E}$, but there has been no mention yet of probabilities. The third thing which we do is to allocate probabilities to each event in $\mathcal{F}$, writing P(A) for the probability of the event A. We shall assume that this can be done in such a way that the probability function P satisfies certain intuitively attractive conditions:

(i) each event A in the event space should have a probability P(A) which lies between 0 and 1;

(ii) the event Ω, that 'something happens', should have probability 1, and the event $\varnothing$, that 'nothing happens', should have probability 0;

(iii) if A and B are disjoint events (so that $A \cap B = \varnothing$) then P($A \cup B$) = P(A) + P(B).

We collect these conditions into a formal definition as follows.

A mapping $P : \mathcal{F} \to \mathbb{R}$ is called a *probability measure* on $(\Omega, \mathcal{F})$ if

(7)
$$P(A) \geq 0 \quad \text{for all} \quad A \in \mathcal{F},$$

(8)
$$P(\Omega) = 1 \quad \text{and} \quad P(\varnothing) = 0,$$

(9) if $A_1, A_2, \ldots$ are disjoint events in $\mathcal{F}$ (so that $A_i \cap A_j = \varnothing$ whenever $i \neq j$) then

$$P\left(\bigcup_{i=1}^{\infty} A_i\right) = \sum_{i=1}^{\infty} P(A_i).$$

We emphasize that a probability measure P on $(\Omega, \mathcal{F})$ is defined only on those subsets of Ω which lie in $\mathcal{F}$. The second part of condition (8) is superfluous in the above definition; to see this, note that $\varnothing$ and Ω are disjoint events with union $\Omega \cup \varnothing = \Omega$ and so

$$P(\Omega) = P(\Omega \cup \varnothing) = P(\Omega) + P(\varnothing) \qquad \text{by (9)}.$$

Condition (9) requires that the probability of the union of a countable† collection of non-overlapping sets is the sum of the individual probabilities.

Example 10   Let Ω be a set and A be a proper subset of Ω (so that $A \neq \varnothing, \Omega$). If $\mathcal{F}$ is the event space $\{\varnothing, A, \Omega \backslash A, \Omega\}$ then all probability measures on $(\Omega, \mathcal{F})$ have the form

$$P(\varnothing) = 0, \qquad \qquad P(A) = p,$$
$$P(\Omega \backslash A) = 1 - p, \qquad P(\Omega) = 1,$$

for some p satisfying $0 \leq p \leq 1$.   □

---

†A set S is called *countable* if it may be put in one–one correspondence with a subset of the natural numbers $\{1, 2, 3, \ldots\}$.

Example 11 Let $\Omega = \{\omega_1, \omega_2, \ldots, \omega_N\}$ be a finite set of exactly N points, and let $\mathcal{F}$ be the power set of Ω. It is easy to check that the function P defined by†

$$P(A) = \frac{1}{N}|A| \qquad \text{for } A \in \mathcal{F}$$

is a probability measure on $(\Omega, \mathcal{F})$. $\qquad\qquad\qquad\qquad\qquad\qquad$ □

Exercises 6. Let $p_1, p_2, \ldots, p_N$ be non-negative numbers such that $p_1 + p_2 + \cdots + p_N = 1$, and let $\Omega = \{\omega_1, \omega_2, \ldots, \omega_N\}$, with $\mathcal{F}$ the power set of Ω, as in Example 11 above. Show that the function Q given by

$$Q(A) = \sum_{i:\omega_i \in A} p_i \qquad \text{for } A \in \mathcal{F},$$

is a probability measure on $(\Omega, \mathcal{F})$. Is Q a probability measure if $\mathcal{F}$ is not the power set of Ω but merely some event space of subsets of Ω?

1.4 Probability spaces

We now combine the previous ideas and define a *probability space* to be a triple $(\Omega, \mathcal{F}, P)$ of objects such that
  (i) Ω is a set,
  (ii) $\mathcal{F}$ is an event space of subsets of Ω,
  (iii) P is a probability measure on $(\Omega, \mathcal{F})$.
There are many elementary consequences of the axioms which underlie this definition, and we describe some of these. Let $(\Omega, \mathcal{F}, P)$ be a probability space.

(12) $\qquad\qquad\qquad\qquad$ If $A, B \in \mathcal{F}$ then‡ $A \backslash B \in \mathcal{F}$.

Proof The complement of $A \backslash B$ equals $(\Omega \backslash A) \cup B$, which is the union of events and is therefore an event. Hence $A \backslash B$ is an event, by (2). □

(13) $\qquad\qquad\qquad\qquad$ If $A_1, A_2, \ldots \in \mathcal{F}$ then $\bigcap_{i=1}^{\infty} A_i \in \mathcal{F}$.

Proof The complement of $\bigcap_{i=1}^{\infty} A_i$ equals $\bigcup_{i=1}^{\infty} (\Omega \backslash A_i)$ which is the union of the complements of events and is therefore an event. Hence the intersection of the A's is an event also, as before. □

(14) $\qquad\qquad\qquad\qquad$ If $A \in \mathcal{F}$ then $P(A) + P(\Omega \backslash A) = 1$.

† The *cardinality* $|A|$ of a set A is the number of points in A.

‡ $A \backslash B = A \cap (\Omega \backslash B)$ is the set of points in A which are not in B.

Proof A and $\Omega \backslash A$ are disjoint events with union Ω, and so

$$1 = P(\Omega) = P(A) + P(\Omega \backslash A). \qquad \square$$

(15)   If $A, B \in \mathscr{F}$ then $P(A \cup B) + P(A \cap B) = P(A) + P(B)$.

Proof   The set A is the union of the disjoint sets $A \backslash B$ and $A \cap B$, and hence

$$P(A) = P(A \backslash B) + P(A \cap B) \qquad \text{by (9)}.$$

A similar remark holds for the set B, giving that

$$\begin{aligned}
P(A) + P(B) &= P(A \backslash B) + 2P(A \cap B) + P(B \backslash A) \\
&= P((A \backslash B) \cup (A \cap B) \cup (B \backslash A)) + P(A \cap B) \qquad \text{by (9)} \\
&= P(A \cup B) + P(A \cap B). \qquad \square
\end{aligned}$$

(16)   If $A, B \in \mathscr{F}$ and $A \subseteq B$ then $P(A) \leq P(B)$.

Proof $P(B) = P(A) + P(B \backslash A) \geq P(A)$. $\qquad \square$

It is often useful to draw a Venn diagram when working with probabilities. For example, to show the formula in (15) above we might draw the diagram in Fig. 1.1, and note that the probability of $A \cup B$ is the sum of $P(A)$ and $P(B)$ minus $P(A \cap B)$, since this latter probability is counted twice in the simple sum $P(A) + P(B)$.

Exercises   In these exercises $(\Omega, \mathscr{F}, P)$ is a probability space.
  7. If $A, B \in \mathscr{F}$, show that

$$P(A \backslash B) = P(A) - P(A \cap B).$$

  8. If $A, B, C \in \mathscr{F}$, show that

$$\begin{aligned}
P(A \cup B \cup C) = P(A) + P(B) + P(C) - P(A \cap B) \\
- P(A \cap C) - P(B \cap C) + P(A \cap B \cap C).
\end{aligned}$$

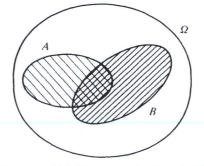

Fig. 1.1   A Venn diagram which illustrates the fact that $P(A \cup B) = P(A) + P(B) - P(A \cap B)$

9. Let A, B, C be three events such that
$$P(A) = \tfrac{5}{10}, \qquad P(B) = \tfrac{7}{10}, \qquad P(C) = \tfrac{6}{10},$$
$$P(A \cap B) = \tfrac{3}{10}, \qquad P(B \cap C) = \tfrac{4}{10}, \qquad P(A \cap C) = \tfrac{2}{10},$$
$$P(A \cap B \cap C) = \tfrac{1}{10}.$$

By drawing a Venn diagram or otherwise, find the probability that exactly two of the events A, B, C occur.
10. A fair coin is tossed 10 times (so that heads appears with probability $\tfrac{1}{2}$ at each toss). Describe the appropriate probability space in detail for the two cases when
   (i) the outcome of every toss is of interest,
   (ii) only the total number of tails is of interest.
   In the first case your event space should have $2^{2^{10}}$ events, but in the second case it should have only 2^{11} events.

1.5 Discrete sample spaces

Let $\mathscr{E}$ be an experiment with probability space $(\Omega, \mathscr{F}, \mathsf{P})$. The structure of this space depends greatly upon whether Ω is a countable set (that is, a finite or countably infinite set) or an uncountable set. If Ω is a countable set then we normally take $\mathscr{F}$ to be the set of *all* subsets of Ω, for the following reason. Suppose that $\Omega = \{\omega_1, \omega_2, \ldots\}$ and, for each $\omega \in \Omega$, we are interested in whether or not this given ω is the actual outcome of $\mathscr{E}$; then we require that each singleton set $\{\omega\}$ belongs to $\mathscr{F}$. Let $A \subseteq \Omega$. Then A is countable (since Ω is countable) and so A may be expressed as the union of the countably many ω's which belong to A, giving that $A = \bigcup_{\omega \in A} \{\omega\} \in \mathscr{F}$ by (3). The probability $P(A)$ of the event A is determined by the collection $\{\mathsf{P}(\{\omega\}) : \omega \in \Omega\}$ of probabilities since, by (9),

$$P(A) = \sum_{\omega \in A} \mathsf{P}(\{\omega\}).$$

We usually write $\mathsf{P}(\omega)$ for the probability $\mathsf{P}(\{\omega\})$ of an event containing only one point in Ω.

Example 17

Equiprobable outcomes. If $\Omega = \{\omega_1, \omega_2, \ldots, \omega_N\}$ and $\mathsf{P}(\omega_i) = \mathsf{P}(\omega_j)$ for all i and j, then $\mathsf{P}(\omega) = N^{-1}$ for all $\omega \in \Omega$, and $P(A) = |A|/N$ for all $A \subseteq \Omega$. $\qquad \square$

Example 18

Random integers. There are "intuitively-clear" statements which are without meaning in probability theory, and here is an example: *if we pick a positive integer at random, then it is an even integer with probability $\tfrac{1}{2}$.* Interpreting "at random" to mean that each positive integer is equally likely to be picked, then this experiment would have probability space $(\Omega, \mathscr{F}, P)$ where

(i) $\Omega = \{1, 2, \ldots\}$,

(ii) $\mathscr{F}$ is the set of all subsets of Ω,

(iii) if $A \subseteq \Omega$ then $P(A) = \sum_{i \in A} P(i) = \pi |A|$ where π is the probability that any given integer, i say, is picked.

However,

$$\text{if } \pi = 0 \text{ then } P(\Omega) = \sum_{i=1}^{\infty} 0 = 0,$$

$$\text{if } \pi > 0 \text{ then } P(\Omega) = \sum_{i=1}^{\infty} \pi = \infty,$$

neither of which is in agreement with the rule that $P(\Omega) = 1$. One possible way of interpreting the italicized statement above is as follows. Fix N, a large positive integer, and let $\mathscr{E}_N$ be the experiment of picking an integer from $\Omega_N = \{1, 2, \ldots, N\}$ at random. The probability that the outcome of $\mathscr{E}_N$ is even is

$$\tfrac{1}{2} \text{ if } N \text{ is even, and } \frac{1}{2}\left(1 - \frac{1}{N}\right) \text{ if } N \text{ is odd,}$$

so that, as $N \to \infty$, the required probability tends to $\tfrac{1}{2}$. Despite this sensible interpretation of the italicized statement, we emphasize that this statement is without meaning in its present form and should be shunned by serious probabilists. $\qquad\qquad\square$

Exercises The most elementary problems in probability theory are those which involve experiments such as the shuffling of cards or the throwing of dice, and these usually give rise to situations in which every possible outcome is equally likely to occur. This is the case of Example 17 above. Such problems usually reduce to the problem of counting the number of ways in which some event may occur, and the following exercises are of this type.

11. Show that if a coin is tossed n times, then there are exactly

$$\binom{n}{k} = \frac{n!}{k!\,(n-k)!}$$

sequences of possible outcomes in which exactly k heads are obtained. If the coin is fair (so that heads and tails are equally likely on each toss), show that the probability of getting at least k heads is

$$\frac{1}{2^n} \sum_{r=k}^{n} \binom{n}{r}.$$

12. We distribute r distinguishable balls into n cells at random, multiple occupancy being permitted. Show that
(i) there are n^r possible arrangements,
(ii) there are $\binom{r}{k}(n-1)^{r-k}$ arrangements in which the first cell contains exactly k balls,

(iii) the probability that the first cell contains exactly k balls is

$$\binom{r}{k}\left(\frac{1}{n}\right)^{k}\left(1-\frac{1}{n}\right)^{r-k}.$$

13. Show that the probability that each of the four players in a game of bridge receives one ace is

$$\frac{24 \cdot 48! \, 13^{4}}{52!} = 0.105. \ldots$$

14. Show that the probability that two given hands in bridge contain k aces between them is

$$\binom{4}{k}\binom{48}{26-k}\Big/\binom{52}{26}.$$

15. Show that the probability that a hand in bridge contains 6 spades, 3 hearts, 2 diamonds, and 2 clubs is

$$\binom{13}{6}\binom{13}{3}\binom{13}{2}^{2}\Big/\binom{52}{13}.$$

16. Which of the following is more probable:
    (i) getting at least one six with 4 throws of a die,
    (ii) getting at least one double six with 24 throws of two dice?
    This is sometimes called 'de Méré's paradox', after the professional gambler Chevalier de Méré who believed these two events to have equal probability.

1.6   Conditional probabilities

Let $\mathscr{E}$ be an experiment with probability space $(\Omega, \mathscr{F}, \mathsf{P})$. We may sometimes possess incomplete information about the actual outcome of $\mathscr{E}$ without knowing this outcome exactly. For example, if we throw a fair die and a friend tells us that an even number is showing, then this information affects all our calculations of probabilities. In general, if A and B are events (that is $A, B \in \mathscr{F}$) and we are given that B occurs, then, in the light of this information, the new probability of A may no longer be $\mathsf{P}(A)$. Clearly, in this new circumstance, A occurs if and only if $A \cap B$ occurs, suggesting that the new probability of A is proportional to $\mathsf{P}(A \cap B)$. We make this chat more formal in a definition.

If $A, B \in \mathscr{F}$ and $\mathsf{P}(B) > 0$, then the (*conditional*) *probability of* A *given* B is denoted by $\mathsf{P}(A \mid B)$ and defined by

(19)
$$\mathsf{P}(A \mid B) = \frac{\mathsf{P}(A \cap B)}{\mathsf{P}(B)}.$$

Note that the constant of proportionality in (19) hs been chosen so that the probability $P(B \mid B)$, that B occurs given that B occurs, satisfies $P(B \mid B) = 1$. We must first check that this definition gives rise to a probability space.

Theorem 1A If $B \in \mathcal{F}$ and $P(B) > 0$ then $(\Omega, \mathcal{F}, Q)$ is a probability space where $Q : \mathcal{F} \to \mathbb{R}$ is defined by $Q(A) = P(A \mid B)$.

Proof We need only check that Q is a probability measure on $(\Omega, \mathcal{F})$. Certainly $Q(A) \geq 0$ for all $A \in \mathcal{F}$ and

$$Q(\Omega) = P(\Omega \mid B) = \frac{P(\Omega \cap B)}{P(B)} = 1$$

and it remains to check that Q satisfies (9). Suppose that $A_1, A_2, \ldots$ are disjoint events in $\mathcal{F}$. Then

$$Q\left(\bigcup_i A_i\right) = \frac{1}{P(B)} P\left(\left(\bigcup_i A_i\right) \cap B\right)$$

$$= \frac{1}{P(B)} P\left(\bigcup_i (A_i \cap B)\right)$$

$$= \frac{1}{P(B)} \sum_i P(A_i \cap B) \qquad \text{since P satisfies (9)}$$

$$= \sum_i Q(A_i). \qquad \square$$

Exercises 17. If $(\Omega, \mathcal{F}, P)$ is a probability space and A, B, C are events, show that

$$P(A \cap B \cap C) = P(A \mid B \cap C)P(B \mid C)P(C)$$

so long as $P(B \cap C) > 0$.
18. Show that

$$P(B \mid A) = P(A \mid B)\frac{P(B)}{P(A)}$$

if $P(A) > 0$ and $P(B) > 0$.
19. Consider the experiment of tossing a fair coin 7 times. Find the probability of getting a prime number of heads given that heads occurs on at least 6 of the tosses.

1.7 Independent events

We call two events A and B 'independent' if the occurrence of one of them does not affect the probability that the other occurs; more formally this requires that, if $P(A), P(B) > 0$, then

(20) $$P(A \mid B) = P(A) \quad \text{and} \quad P(B \mid A) = P(B).$$

Writing $P(A \mid B) = P(A \cap B)/P(B)$, we see that the following definition is appropriate.

Events A and B of a probability space $(\Omega, \mathscr{F}, P)$ are called
independent if

(21)
$$P(A \cap B) = P(A)P(B)$$

and *dependent* otherwise.

This definition is slightly more general than (20) since it allows A
and B to have zero probability. It is easily generalized to more than
two events.

A family $\mathscr{A} = (A_i : i \in I)$ of events is called *independent* if, for all
finite subsets J of I,

(22)
$$P\left(\bigcap_{i \in J} A_i\right) = \prod_{i \in J} P(A_i).$$

The family $\mathscr{A}$ is called *pairwise independent* if (22) holds whenever
$|J| = 2$.

Thus, three events A, B, C are independent if and only if the
following equalities hold:

$$P(A \cap B \cap C) = P(A)P(B)P(C), \qquad P(A \cap B) = P(A)P(B),$$
$$P(A \cap C) = P(A)P(C), \qquad P(B \cap C) = P(B)P(C).$$

There are families of events which are pairwise independent but not
independent.

Example 23 Suppose that we throw a fair four-sided die (you may think of this as
a square die thrown in a two-dimensional universe). Then we may
take $\Omega = \{1, 2, 3, 4\}$ where each $\omega \in \Omega$ is equally likely to occur. The
events $A = \{1, 2\}$, $B = \{1, 3\}$, $C = \{1, 4\}$ are pairwise independent
but not independent.   □

Exercises 20. Let A and B be events satisfying $P(A)$, $P(B) > 0$, and such that
$P(A \mid B) = P(A)$. Show that $P(B \mid A) = P(B)$.
21. If A and B are events which are disjoint and independent, what can be
said about the probabilities of A and B?
22. Show that events A and B are independent if and only if A and $\Omega \backslash B$ are
independent.
23. Show that events $A_1, A_2, \ldots, A_m$ are independent if and only if $\Omega \backslash A_1$,
$\Omega \backslash A_2, \ldots, \Omega \backslash A_m$ are independent.
24. If $A_1, A_2, \ldots, A_m$ are independent and $P(A_i) = p$ for $i = 1, 2, \ldots, m$,
find the probability that
   (i) none of the A's occur,
   (ii) an even number of the A's occur.
25. On your desk there is a very special die which has a prime number p of
faces, and you throw this die once. Show that no two events A and B can
be independent unless either A or B is the whole sample space or the
empty set.

1.8 The partition theorem

Let $(\Omega, \mathscr{F}, P)$ be a probability space. A *partition* of Ω is a collection $\{B_i : i \in I\}$ of disjoint events (so that $B_i \in \mathscr{F}$ for each i and $B_i \cap B_j = \emptyset$ if $i \neq j$) with union $\bigcup_i B_i = \Omega$. The following *partition theorem* is extremely useful.

Theorem 1B   *If $\{B_1, B_2, \ldots\}$ is a partition of Ω such that $P(B_i) > 0$ for each i, then*

$$P(A) = \sum_i P(A \mid B_i)P(B_i) \quad \text{for all} \quad A \in \mathscr{F}.$$

This theorem has several other fancy names, such as 'the theorem of total probability'; it is closely related to the so-called 'Bayes formula'.

Proof

$$P(A) = P\left(A \cap \left(\bigcup_i B_i\right)\right)$$

$$= P\left(\bigcup_i (A \cap B_i)\right)$$

$$= \sum_i P(A \cap B_i) \qquad \text{by (9)}$$

$$= \sum_i P(A \mid B_i)P(B_i) \qquad \text{by (19).} \qquad \square$$

Here is an example of this theorem in action.

Example 24   Tomorrow there will be either rain or snow but not both; the probability of rain is $\frac{2}{5}$ and the probability of snow is $\frac{3}{5}$. If it rains then the probability that I will be late for my lecture is $\frac{1}{5}$, while the corresponding probability in the event of snow is $\frac{3}{5}$. What is the probability that I will be late?

Solution   Let A be the event that I am late and B be the event that it rains. Then the pair B, B^c is a partition of the sample space (since one or other of them must occur). We apply Theorem 1B to find that

$$P(A) = P(A \mid B)P(B) + P(A \mid B^c)P(B^c)$$
$$= \tfrac{1}{5} \cdot \tfrac{2}{5} + \tfrac{3}{5} \cdot \tfrac{3}{5} = \tfrac{11}{25}. \qquad \square$$

Exercises   26. Here is a routine problem about balls in urns. You are presented with two urns. Urn I contains 3 white and 4 black balls and Urn II contains 2 white and 6 black balls. You pick a ball randomly from Urn I and place it in Urn II. Next you pick a ball randomly from Urn II. What is the probability that this ball is black?

27. A biased coin shows heads with probability $p = 1 - q$ whenever it is

tossed. Let u_n be the probability that, in n tosses, no pair of heads occur successively. Show that for $n \geq 1$

$$u_{n+2} = qu_{n+1} + pqu_n,$$

and find u_n by the usual method (described in the appendix) if $p = \frac{2}{3}$. (Hint: use Theorem 1B with B_i the event that the first $i - 1$ tosses yield heads and the ith yields tails.

1.9 Probability measures are continuous

There is a certain property of probability measures which will be crucially important later, and we describe this next. Too great an emphasis should not be laid on this property at this stage, and therefore we strongly recommend to the reader that he omit this section at the first reading.

A sequence $A_1, A_2, \ldots$ of events in a probability space $(\Omega, \mathcal{F}, P)$ is called *increasing* if

$$A_1 \subseteq A_2 \subseteq \cdots \subseteq A_{n-1} \subseteq A_n \subseteq A_{n+1} \subseteq \cdots.$$

The union

$$A = \bigcup_{i=1}^{\infty} A_i$$

of such a sequence is called the *limit* of the sequence, and it is an elementary consequence of the axioms for an event space that A is an event. It is perhaps not surprising that the probability $P(A)$ of A may be expressed as the limit $\lim_{n \to \infty} P(A_n)$ of the probabilities of the A's.

Theorem 1C *Let $(\Omega, \mathcal{F}, P)$ be a probability space. If $A_1, A_2, \ldots$ is an increasing sequence of events in $\mathcal{F}$ with limit A, then*

$$P(A) = \lim_{n \to \infty} P(A_n).$$

We precede the proof of the theorem with an application.

Example 25 It is intuitively clear that the chance of obtaining no heads in an infinite set of tosses of a fair coin is 0. A rigorous proof goes as follows. Let A_n be the event that the first n tosses of the coin yield at least one head. Then

$$A_n \subseteq A_{n+1} \quad \text{for } n = 1, 2, \ldots,$$

so that the A's form an increasing sequence; the limit set A is the event that heads occurs sooner or later. From Theorem 1C,

$$P(A) = \lim_{n \to \infty} P(A_n);$$

however, $P(A_n) = 1 - (\frac{1}{2})^n$ and so $P(A_n) \to 1$ as $n \to \infty$. Thus $P(A) = 1$, giving that the probability $P(\Omega \backslash A)$, that heads never appears, equals 0. □

Proof of   Let $B_i = A_i \backslash A_{i-1}$ for $i = 2, 3, \ldots$. Then
Theorem

$$A = A_1 \cup B_2 \cup B_3 \cup \cdots$$

is the union of disjoint events in $\mathscr{F}$ (draw a Venn diagram to make this clear). By (9),

$$P(A) = P(A_1) + P(B_2) + P(B_3) + \cdots$$

$$= P(A_1) + \lim_{n \to \infty} \sum_{k=2}^{n} P(B_k).$$

However,

$$P(B_i) = P(A_i) - P(A_{i-1}) \quad \text{for } i = 2, 3, \ldots,$$

and so
$$P(A) = P(A_1) + \lim_{n \to \infty} \sum_{k=2}^{n} [P(A_k) - P(A_{k-1})]$$

$$= \lim_{n \to \infty} P(A_n)$$

as required, since the last sum collapses. □

The conclusion of Theorem 1C is expressed in terms of *increasing* sequences of events, but the corresponding result about *decreasing* sequences is valid too: if $B_1, B_2, \ldots$ is a sequence of events in $\mathscr{F}$ such that $B_i \supseteq B_{i+1}$ for $i = 1, 2, \ldots$, then $P(B_n) \to P(B)$ as $n \to \infty$, where $B = \bigcap_{i=1}^{\infty} B_i$ is the limit of the B_i's as $i \to \infty$. The easiest way to deduce this is to set $A_i = \Omega \backslash B_i$ in the theorem.

1.10 Worked problems

Example   A fair six-sided die is thrown twice (when applied to such objects as
26   dice or coins, the adjectives 'fair' and 'unbiased' imply that each possible outcome has equal probability of occurring).
(a) Write down the probability space of this experiment.
(b) Let B be the event that the first number thrown is no larger than 3, and let C be the event that the sum of the two numbers thrown equals 6. Find the probabilities of B and C, and the conditional probabilities of C given B, and of B given C.

Solution   The probability space of this experiment is the triple $(\Omega, \mathscr{F}, P)$ where
(i) $\Omega = \{(i, j) : i, j = 1, 2, \ldots, 6\}$ is the set of all ordered pairs of integers between 1 and 6,

(ii) $\mathcal{F}$ is the set of all subsets of Ω,

(iii) each point in Ω has equal probability, so that

$$P((i, j)) = \tfrac{1}{36} \quad \text{for } i, j = 1, 2, \ldots, 6,$$

and, more generally,

$$P(A) = \tfrac{1}{36} |A| \quad \text{for each } A \subseteq \Omega.$$

The events B and C are subsets of Ω given by

$$B = \{(i, j) : i = 1, 2, 3 \text{ and } j = 1, 2, \ldots, 6\},$$
$$C = \{(i, j) : i + j = 6 \text{ and } i, j = 1, 2, \ldots, 6\}.$$

B contains $3 \times 6 = 18$ ordered pairs, and C contains 5 ordered pairs, giving that

$$P(B) = \tfrac{18}{36} = \tfrac{1}{2}, \qquad P(C) = \tfrac{5}{36}.$$

Finally, $B \cap C$ is given by

$$B \cap C = \{(1, 5), (2, 4), (3, 3)\}$$

containing just 3 ordered pairs, so that

$$P(C \mid B) = \frac{P(C \cap B)}{P(B)} = \tfrac{3}{36} / \tfrac{18}{36} = \tfrac{1}{6},$$

and

$$P(B \mid C) = \frac{P(B \cap C)}{P(C)} = \tfrac{3}{36} / \tfrac{5}{36} = \tfrac{3}{5}. \qquad \square$$

Example 27   You are travelling in a train with your sister. Neither of you has a valid ticket and the inspector has caught you both. He is authorized to administer a special punishment for this offence. He holds a box containing nine apparently identical chocolates, but three of these are contaminated with a deadly poison. He makes each of you, in turn, choose and immediately eat a single chocolate.

(a) If you choose before your sister, what is the probability that you survive?

(b) If you choose first and survive, what is the probability that your sister survives?

(c) If you choose first and die, what is the probability that your sister survives?

(d) Is it in your best interests to persuade your sister to choose first?

(e) If you choose first, what is the probability that you survive, given that your sister survives?

Solution   Let A be the event that the first chocolate picked is not poisoned, and let B be the event that the second chocolate picked is not poisoned. Elementary calculations, if you are allowed the time to perform them, would show that

$$P(A) = \tfrac{6}{9}, \qquad P(B \mid A) = \tfrac{5}{8}, \qquad P(B \mid A^c) = \tfrac{6}{8},$$

giving by the partition theorem 1B that

$$P(B) = P(B \mid A)P(A) + P(B \mid A^c)P(A^c)$$
$$= \tfrac{5}{8} \cdot \tfrac{6}{9} + \tfrac{6}{8} \cdot (1 - \tfrac{6}{9}) = \tfrac{2}{3}.$$

Hence $P(A) = P(B)$, so that the only reward of choosing second is to increase your life expectancy by a few seconds.

The final question (e) seems to be the wrong way round in time, since your sister chooses her chocolate *after* you. The way to answer such a question is to reverse the conditioning as follows:

(28)
$$P(A \mid B) = \frac{P(A \cap B)}{P(B)} = P(B \mid A)\frac{P(A)}{P(B)},$$

and hence

$$P(A \mid B) = \tfrac{5}{8} \cdot \tfrac{6}{9} / \tfrac{2}{3} = \tfrac{5}{8}.$$

We note that $P(A \mid B) = P(B \mid A)$, in agreement with our earlier observation that the order in which you and your sister pick from the box is irrelevant to your chances of survival.   □

1.11 Problems

1. A fair die is thrown n times. Show that the probability that there are an even number of sixes is

$$\tfrac{1}{2}[1 + (\tfrac{2}{3})^n].$$

   For the purpose of this question, 0 is an even number.
2. Does there exist an event space which contains just six events?
3. Prove Boole's inequality:

$$P\left(\bigcup_{i=1}^{n} A_i\right) \le \sum_{i=1}^{n} P(A_i).$$

4. Two fair dice are thrown. Let A be the event that the first shows an odd number, B be the event that the second shows an even number, and C be the event that either both are odd or both are even. Show that A, B, C are pairwise independent but not independent.
5. Urn I contains 4 white and 3 black balls, and Urn II contains 3 white and 7 black balls. An urn is selected at random and a ball is picked from it. What is the probability that this ball is black? If this ball is white, what is the probability that Urn I was selected?
6. A single card is removed at random from a deck of 52 cards. From the

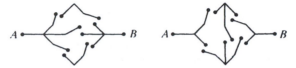

Fig. 1.2 Two electrical circuits incorporating switches

remainder we draw two cards at random and find that they are both spades. What is the probability that the first card removed was also a spade?

7. Two people toss a fair coin n times each. Show that the probability they throw equal numbers of heads is

$$\binom{2n}{n}\left(\frac{1}{2}\right)^{2n}.$$

8. In the circuits in Fig. 1.2 each switch is closed with probability p, independently of all other switches. For each circuit, find the probability that a flow of current is possible between A and B.

9. Show that if u_n is the probability that n tosses of a fair coin contain no run of 4 heads, then for $n \geq 4$

$$u_n = \tfrac{1}{2}u_{n-1} + \tfrac{1}{4}u_{n-2} + \tfrac{1}{8}u_{n-3} + \tfrac{1}{16}u_{n-4}.$$

Use this difference equation to show that, in 8 throws, the probability of no such run is $\frac{208}{256}$.

*10. Any number $\omega \in [0, 1]$ has a decimal expansion

$$\omega = 0 \cdot x_1 x_2 \ldots,$$

and we write $f_k(\omega, n)$ for the proportion of times that the integer k appears in the first n digits in this expansion. We call ω a *normal number* if

$$f_k(\omega, n) \to \tfrac{1}{10} \quad \text{as} \quad n \to \infty$$

for $k = 0, 1, 2, \ldots, 9$. On intuitive grounds we may expect that most numbers $\omega \in [0, 1]$ are normal numbers, and Borel proved that this is indeed true. It is quite another matter to exhibit specific normal numbers. It is known that the number

$$0.1234567891011121314\ldots$$

is normal; prove this. It is an unsolved problem of mathematics to show that $e - 2$ and $\pi - 3$ are normal numbers also.

11. A square board is divided into 16 equal squares by lines drawn parallel to its sides. A counter is placed at random on one of these squares and is then moved n times. At each of these moves it can be transferred to any neighbouring square, horizontally, vertically, or diagonally, all such moves being equally likely.

Let c_n be the probability that a particular corner site is occupied after n such independent moves, and let the corresponding probabilities for an intermediate site at the side of the board and for a site in the middle of the board be s_n and m_n respectively. Show that

$$4c_n + 8s_n + 4m_n = 1, \qquad n = 0, 1, 2, \ldots,$$

and that

$$c_n = \tfrac{2}{5}s_{n-1} + \tfrac{1}{8}m_{n-1}, \qquad n = 1, 2, \dots .$$

Find two other relations for s_n and m_n in terms of c_{n-1}, s_{n-1}, and m_{n-1} and hence find c_n, s_n, and m_n.

(Oxford 1974M)

12. (a) Let $P(A)$ denote the probability of the occurrence of an event A. Prove carefully, for events $A_1, A_2, \dots, A_n$, that

$$P\left(\bigcup_{1 \le i \le n} A_i\right) = \sum_{1 \le i \le n} P(A_i) - \sum_{1 \le i < j \le n} P(A_i \cap A_j) + \cdots$$
$$+ (-1)^{n+1} P\left(\bigcap_{1 \le i \le n} A_i\right).$$

(b) One evening a bemused lodge-porter tried to hang n keys on their n hooks, but only managed to hang them independently and at random. There was no limit to the number of keys which could be hung on any hook. Otherwise, or by using (a), find an expression for the probability that at least one key was hung on its own hook. The following morning the porter was rebuked by the Bursar, so that in the evening he was careful to hang only one key on each hook. But he still only managed to hang them independently and at random. Find an expression for the probability that no key was then hung on its own hook.

Find the limits of both expressions as n tends to infinity.

(You may assume that

$$e^x = \sum_{r=0}^{\infty} \frac{x^r}{r!} = \lim_{N \to \infty} \left(1 + \frac{x}{N}\right)^N$$

for all real x.)

(Oxford 1978M)

13. Two identical decks of cards, each containing N cards, are shuffled randomly. We say that a *k-matching* occurs if the two decks agree in exactly k places. Show that the probability that there is a k-matching is

$$\pi_k = \frac{1}{k!} \left(1 - \frac{1}{1!} + \frac{1}{2!} - \frac{1}{3!} + \cdots + \frac{(-1)^{N-k}}{(N-k)!}\right)$$

for $k = 0, 1, 2, \dots, N-1, N$. We note that $\pi_k \approx (k!\,e)^{-1}$ for large N and fixed k. Such matching probabilities are used in testing departures from randomness in circumstances such as psychological tests and wine-tasting competitions. (The convention is that $0! = 1$.)

14. The buses which stop at the end of my road do not keep to the timetable. They should run every quarter hour, at 08.30, 08.45, 09.00, ..., but in fact each bus is either five minutes early or five minutes late, the two possibilities being equally probable and different buses being independent. Other people arrive at the stop in such a way that, t minutes after the departure of one bus, the probability that no one is waiting for the next one is $e^{-t/5}$. What is the probability that no one is waiting at 09.00? One day I come to the stop at 09.00 and find no one there; show that the chances are more than four to one that I have missed the nine o'clock bus.

(You may use the good approximation $e^3 \approx 20$.)

(Oxford 1977M)

15. A coin is tossed repeatedly; on each toss a head is shown with probability p, a tail with probability $1 - p$. All tosses are mutually

independent. Let E denote the event that the first run of r successive heads occurs earlier than the first run of s successive tails. Let A denote the outcome of the first toss. Show that

$$P(E \mid A = \text{head}) = p^{r-1} + (1 - p^{r-1})P(E \mid A = \text{tail}).$$

Find a similar expression for $P(E \mid A = \text{tail})$ and hence find $P(E)$.

(Oxford 1981M)

16. Try Eddington's Controversy: If A, B, C, D each speak the truth once in three times (independently) and A claims that B denies that C declares that D is a liar, what is the probability that D was speaking the truth?

*17. Show that the axiom that P is countably additive is equivalent to the axiom that P is finitely additive and continuous. That is to say, let Ω be a set and $\mathscr{F}$ an event space of subsets of Ω. If P is a mapping from $\mathscr{F}$ into $[0, 1]$ satisfying
    (i) $P(\Omega) = 1$, $P(\emptyset) = 0$,
    (ii) if $A, B \in \mathscr{F}$ and $A \cap B = \emptyset$ then $P(A \cup B) = P(A) + P(B)$,
    (iii) if $A_1, A_2, \ldots \in \mathscr{F}$ and $A_i \subseteq A_{i+1}$ for $i = 1, 2, \ldots$, then

$$P(A) = \lim_{i \to \infty} P(A_i)$$

where $A = \bigcup_{i=1}^{\infty} A_i$,
then P satisfies $P(\bigcup_{i=1}^{\infty} A_i) = \sum_i P(A_i)$ for all sequences $A_1, A_2, \ldots$ of disjoint events.

18. In a piece of electronic equipment, a gate, which may be open or closed, has a random method of operation. Every millisecond it may change its state; the changes occur in the following manner:
    (i) if it is open it remains open with probability $1 - \alpha$ and becomes closed with probability α,
    (ii) if it is closed it remains so with probability $1 - \beta$ and becomes open with probability β.
    If $0 < \alpha < 1$ and $0 < \beta < 1$, and θ_n is the probability that the gate is closed after n milliseconds, derive a recurrence relation for θ_n, showing clearly how your result is obtained. Let θ_0 be the probability that the gate is closed initially. Solve your recurrence relation to give θ_n in terms of α, β, θ_0, and n. Make sure that you justify your solution. Hence find $\lim_{n \to \infty} \theta_n$ and explain intuitively why this does not depend on θ_0.
    Find θ_n in each of the four cases $\alpha = 0$, $\beta = 0$; $\alpha = 0$, $\beta = 1$; $\alpha = 1$, $\beta = 0$; and $\alpha = 1$, $\beta = 1$. Explain the behaviour of the system in each case. (Bristol 1982)

2

Discrete random variables

2.1 Probability mass functions

Given a probability space $(\Omega, \mathcal{F}, P)$, we are often interested in situations involving some real-valued function X on Ω. For example, let $\mathcal{E}$ be the experiment of throwing a fair die once, so that $\Omega = \{1, 2, 3, 4, 5, 6\}$, and suppose that we gamble on the outcome of $\mathcal{E}$ in such a way that our profit is

-1 if the outcome is 1, 2, or 3,

0 if the outcome is 4,

2 if the outcome is 5 or 6,

where negative profits are positive losses. If the outcome is ω then our profit is $X(\omega)$ where $X : \Omega \rightarrow \mathbb{R}$ is defined by

$$X(1) = X(2) = X(3) = -1, \qquad X(4) = 0, \qquad X(5) = X(6) = 2.$$

The mapping X is an example of a 'discrete random variable'.

More formally, a *discrete random variable* X on the probability space $(\Omega, \mathcal{F}, P)$ is defined to be a mapping $X : \Omega \rightarrow \mathbb{R}$ such that

(1)   the image† $X(\Omega)$ of Ω under X is a countable subset of $\mathbb{R}$,

(2) $\{\omega \in \Omega : X(\omega) = x\} \in \mathcal{F}$ for all $x \in \mathbb{R}$.

The word 'discrete' here refers to the condition that X takes only countably many values in $\mathbb{R}$. Condition (2) is obscure at first sight, and the point here is as follows. A discrete random variable X takes values in $\mathbb{R}$, but we cannot predict the actual value of X with certainty since the underlying experiment $\mathcal{E}$ involves chance; instead, we would like to assess the probability that X takes any given value, x say. To this end we note that X takes the value x if and only if the result of $\mathcal{E}$ lies in that subset of Ω which is mapped into x, namely $X^{-1}(x) = \{\omega \in \Omega : X(\omega) = x\}$. Condition (2) postulates that all such subsets are events, in that they belong to $\mathcal{F}$, and are therefore assigned probabilities by P.

The most interesting things about a discrete random variable are

---

† If $X : \Omega \rightarrow \mathbb{R}$ and $A \subseteq \Omega$, the *image* $X(A)$ of A is the set $\{X(\omega) : \omega \in A\}$ of values taken by X on A.

the values which it may take and the probabilities associated with these values.

If X is a discrete random variable on the probability space $(\Omega, \mathcal{F}, \mathsf{P})$, then the *image* Im X of X is the image of Ω under X; the image of X is the set of values taken by X. The (*probability*) *mass function* p_X of X is the function which maps $\mathbb{R}$ into $[0, 1]$ defined by

(3)
$$p_X(x) = \mathsf{P}(\{\omega \in \Omega : X(\omega) = x\}).$$

Thus, $p_X(x)$ is the probability that the mapping X takes the value x. We usually abbreviate this probability to '$\mathsf{P}(X = x)$'. Note that Im X is countable for any discrete random variable X, and

(4)
$$p_X(x) = 0 \text{ if } x \notin \text{Im } X,$$

(5)
$$\sum_{x \in \text{Im } X} p_X(x) = \mathsf{P}\left(\bigcup_{x \in \text{Im } X} \{\omega \in \Omega : X(\omega) = x\}\right) \quad \text{by (1.9)}$$
$$= \mathsf{P}(\Omega) = 1.$$

Equation (5) is sometimes written as

$$\sum_{x \in \mathbb{R}} p_X(x) = 1,$$

in the light of the fact that only countably many values of x make non-zero contributions to this sum. Condition (5) essentially characterizes mass functions of discrete random variables in the sense of the following theorem.

Theorem 2A   *If* $S = \{s_i : i \in I\}$ *is a countable set of distinct real numbers and* $\{\pi_i : i \in I\}$ *is a collection of numbers satisfying*

$$\pi_i \geq 0 \text{ for all } i \in I, \text{ and } \sum_{i \in I} \pi_i = 1,$$

then there exists a probability space $(\Omega, \mathcal{F}, \mathsf{P})$ *and a discrete random variable* X *on* $(\Omega, \mathcal{F}, \mathsf{P})$ *such that the probability mass function* p_X *of* X *is given by*

$$p_X(s_i) = \pi_i \qquad \text{for all } i \in I,$$
$$p_X(s) = 0 \qquad \text{if } s \notin S.$$

Proof   Take $\Omega = S$, $\mathcal{F}$ to be the set of all subsets of Ω and

$$\mathsf{P}(A) = \sum_{i : s_i \in A} \pi_i \qquad \text{for all } A \in \mathcal{F}.$$

Finally define $X : \Omega \to \mathbb{R}$ by $X(\omega) = \omega$ for all $\omega \in \Omega$. $\square$

This theorem is very useful, since for many purposes it allows us to forget about sample spaces, event spaces and probability measures; we need only say 'let X be a random variable taking the value s_i with probability π_i, for $i \in I$' and we can be sure that such a random variable exists without having to construct it explicitly.

In the next section we present a list of some of the most common types of discrete random variables.

Exercises

1. If X and Y are discrete random variables on the probability space $(\Omega, \mathscr{F}, \mathsf{P})$ show that U and V are discrete random variables on this space also, where

$$U(\omega) = X(\omega) + Y(\omega), \qquad V(\omega) = X(\omega)Y(\omega),$$

for all $\omega \in \Omega$.

2. Show that if $\mathscr{F}$ is the power set of Ω then all functions which map Ω into a countable subset of $\mathbb{R}$ are discrete random variables.

3. If E is an event of the probability space $(\Omega, \mathscr{F}, \mathsf{P})$ show that the *indicator function* of E, defined to be the function χ_E on Ω given by

$$\chi_E(\omega) = \begin{cases} 1 & \text{if } \omega \in E, \\ 0 & \text{if } \omega \notin E, \end{cases}$$

is a discrete random variable.

4. Let $(\Omega, \mathscr{F}, \mathsf{P})$ be a probability space in which

$$\Omega = \{1, 2, 3, 4, 5, 6\}, \qquad \mathscr{F} = \{\varnothing, \{2, 4, 6\}, \{1, 3, 5\}, \Omega\},$$

and let U, V, W be functions on Ω defined by

$$U(\omega) = \omega, \qquad V(\omega) = \begin{cases} 1 & \text{if } \omega \text{ is even,} \\ 0 & \text{if } \omega \text{ is odd,} \end{cases} \qquad W(\omega) = \omega^2,$$

for all $\omega \in \Omega$. Determine which of U, V, W are discrete random variables on the probability space.

5. For what value of c is the function p, defined by

$$p(k) = \begin{cases} \dfrac{c}{k(k+1)} & \text{if } k = 1, 2, \ldots, \\ 0 & \text{otherwise,} \end{cases}$$

a mass function?

2.2 Examples

There are certain classes of discrete random variables which occur very frequently, and we list some of these. Throughout this section n is a positive integer, p is a number in $[0, 1]$, and $q = 1 - p$. We never describe the underlying probability space.

Bernoulli distribution. We say that the discrete random variable X

has the Bernoulli distribution with parameter p if the image of X is $\{0, 1\}$, so that X takes the values 0 and 1 only, and

(6)
$$P(X = 0) = q, \qquad P(X = 1) = p.$$

Of course, (6) asserts that the mass function of X is given by $p_X(0) = q$, $p_X(1) = p$, $p_X(x) = 0$ if $x \neq 0, 1$.

Binomial distribution. We say that X has the binomial distribution with parameters n and p if X has image $\{0, 1, \ldots, n\}$ and

(7)
$$P(X = k) = \binom{n}{k} p^k q^{n-k} \quad \text{for } k = 0, 1, 2, \ldots, n.$$

Note that (7) gives rise to a mass function satisfying (5) since, by the binomial theorem,

$$\sum_{k=0}^{n} \binom{n}{k} p^k q^{n-k} = (p + q)^n = 1.$$

Poisson distribution. We say that X has the Poisson distribution with parameter $\lambda \ (>0)$ if X has image $\{0, 1, 2, \ldots\}$ and

(8)
$$P(X = k) = \frac{1}{k!} \lambda^k e^{-\lambda} \quad \text{for } k = 0, 1, 2, \ldots.$$

Again, this gives rise to a mass function since

$$\sum_{k=0}^{\infty} \frac{1}{k!} \lambda^k e^{-\lambda} = e^{-\lambda} \sum_{k=0}^{\infty} \frac{1}{k!} \lambda^k = e^{-\lambda} e^{\lambda} = 1.$$

Geometric distribution. We say that X has the geometric distribution with parameter $p \ (>0)$ if X has image $\{1, 2, 3, \ldots\}$ and

(9)
$$P(X = k) = pq^{k-1} \quad \text{for } k = 1, 2, 3, \ldots.$$

As before, note that

$$\sum_{k=1}^{\infty} pq^{k-1} = \frac{p}{1 - q} = 1.$$

Negative binomial distribution. We say that X has the negative binomial distribution with parameters n and $p \ (>0)$ if X has image $\{n, n+1, n+2, \ldots\}$ and

(10)
$$P(X = k) = \binom{k-1}{n-1} p^n q^{k-n} \quad \text{for } k = n, n+1, n+2, \ldots.$$

As before, note that

$$\sum_{k=n}^{\infty} \binom{k-1}{n-1} p^n q^{k-n} = p^n \sum_{l=0}^{\infty} \binom{n+l-1}{l} q^l \qquad \text{where } l = k - n$$

$$= p^n \sum_{l=0}^{\infty} \binom{-n}{l}(-q)^l$$

$$= p^n(1-q)^{-n} = 1,$$

using the binomial expansion of $(1-q)^{-n}$.

Example 11

Here is an example of some of the above distributions in action. Suppose that a coin is tossed n times and there is probability p that heads appears on each toss. Representing heads by H and tails by T, the sample space is the set Ω of all ordered sequences of length n containing H's and T's, where the kth entry of such a sequence represents the result of the kth toss. Ω is finite, and we take $\mathscr{F}$ to be the set of all subsets of Ω. For each $\omega \in \Omega$ we define the probability that ω is the actual outcome by

$$P(\omega) = p^{h(\omega)} q^{t(\omega)}$$

where $h(\omega)$ is the number of heads in ω and $t(\omega) = n - h(\omega)$ is the number of tails in ω. Similarly, for any $A \in \mathscr{F}$,

$$P(A) = \sum_{\omega \in A} P(\omega).$$

For $i = 1, 2, \ldots, n$, we define the discrete random variable X_i by

$$X_i(\omega) = \begin{cases} 1 & \text{if the } i\text{th entry in } \omega \text{ is H,} \\ 0 & \text{if the } i\text{th entry in } \omega \text{ is T.} \end{cases}$$

Then each X_i has image $\{0, 1\}$ and mass function given by

$$P(X_i = 0) = P(\{\omega \in \Omega : \omega_i = T\})$$

where ω_i is the ith entry in ω. Thus

$$P(X_i = 0) = \sum_{\omega \,:\, \omega_i = T} p^{h(\omega)} q^{n - h(\omega)}$$

$$= \sum_{h=0}^{n-1} \sum_{\substack{\omega \,:\, \omega_i = T \\ h(\omega) = h}} p^h q^{n-h}$$

$$= \sum_{h=0}^{n-1} \binom{n-1}{h} p^h q^{n-h}$$

$$= q(p + q)^{n-1} = q$$

and
$$P(X_i = 1) = 1 - P(X_i = 0) = p.$$

Hence each X_i has the Bernoulli distribution with parameter p. We have derived this fact in a very cumbersome manner, but we believe these details to be instructive.

Let
$$S_n = X_1 + \cdots + X_n;$$

more formally, $S_n(\omega) = X_1(\omega) + \cdots + X_n(\omega)$. Clearly S_n is the total number of heads which occur, and S_n takes values in $\{0, 1, \ldots, n\}$ since each X_i equals 0 or 1. Also, for $k = 0, 1, \ldots, n$, we have that

(12)
$$P(S_n = k) = P(\{\omega \in \Omega : h(\omega) = k\})$$
$$= \sum_{\omega \, : \, h(\omega) = k} P(\omega)$$
$$= \binom{n}{k} p^k q^{n-k},$$

and so S_n has the binomial distribution with parameters n and p.

If n is very large and p is very small but np is a 'reasonable size' ($np = \lambda$, say) then the distribution of S_n may be approximated by the Poisson distribution with parameter λ, as follows. For *fixed* $k \geq 0$, write $p = \lambda n^{-1}$ and suppose that n is large to find that

(13)
$$P(S_n = k) = \binom{n}{k} p^k (1-p)^{n-k}$$
$$\approx \frac{n^k}{k!} \left(\frac{\lambda}{n}\right)^k \left(1 - \frac{\lambda}{n}\right)^n \left(1 - \frac{\lambda}{n}\right)^{-k}$$
$$\approx \frac{1}{k!} \lambda^k e^{-\lambda}.$$

This approximation may be useful in practice. For example, consider a single page of the Guardian newspaper containing, say, 10^6 characters, and suppose that the typesetter flips a coin before setting each character and then deliberately mis-sets this character whenever the coin comes up heads. If the coin comes up heads with probability 10^{-5} on each flip, then this is equivalent to taking $n = 10^6$ and $p = 10^{-5}$ in the above example, giving that the number S_n of deliberate mistakes has the binomial distribution with parameters 10^6 and 10^{-5}. It may be easier (and not too inaccurate) to use (13) rather than (12) to calculate probabilities. In this case $\lambda = np = 10$ and so, for example,
$$P(S_n = 10) \approx \frac{1}{10!} (10e^{-1})^{10} \approx 0.125. \qquad \square$$

Example 14 Suppose that we toss the coin of the previous example until the first head turns up, and then we stop. The sample space now is

$$\Omega = \{H, TH, T^2H, \ldots\} \cup \{T^\infty\},$$

where T^kH represents the outcome of k tails followed by a head, and T^∞ represents an infinite sequence of tails with no head. As before $\mathcal{F}$ is the set of all subsets of Ω and P is given by the observation that

$$\mathsf{P}(T^kH) = pq^k \qquad \text{for } k = 0, 1, 2, \ldots,$$

$$\mathsf{P}(T^\infty) = \begin{cases} 1 & \text{if } p = 0, \\ 0 & \text{if } p > 0. \end{cases}$$

Let Y be the total number of tosses in this experiment (so that $Y(T^kH) = k+1$ for $0 \leq k < \infty$ and $Y(T^\infty) = \infty$). If $p > 0$ then

$$\mathsf{P}(Y = k) = \mathsf{P}(T^{k-1}H) = pq^{k-1} \qquad \text{for } k = 1, 2, \ldots,$$

showing that Y has the geometric distribution with parameter p. □

Example 15 If we carry on tossing the coin in the previous example until the nth head has turned up then a similar argument shows that, if $p > 0$, the total number of tosses required has the negative binomial distribution with parameters n and p. □

Exercises 6. If X is a discrete random variable having the Poisson distribution with parameter λ, show that the probability that X is even is $e^{-\lambda} \cosh \lambda$.
7. If X is a discrete random variable having the geometric distribution with parameter p, show that the probability that X is greater than k is $(1 - p)^k$.

2.3 Functions of discrete random variables

Suppose that X is a discrete random variable on the probability space $(\Omega, \mathcal{F}, \mathsf{P})$ and that $g : \mathbb{R} \to \mathbb{R}$. Then it is easy to check that $Y = g(X)$ is a discrete random variable on $(\Omega, \mathcal{F}, \mathsf{P})$ also, defined by

$$Y(\omega) = g(X(\omega)) \qquad \text{for all } \omega \in \Omega.$$

Simple examples are

(16) $$\text{if } g(x) = ax + b \text{ then } g(X) = aX + b,$$

(17) $$\text{if } g(x) = cx^2 \text{ then } g(X) = cX^2.$$

If $Y = g(X)$ then the mass function of Y is given by

(18) $$p_Y(y) = \mathsf{P}(Y = y) = \mathsf{P}(g(X) = y)$$
$$= \mathsf{P}(X \in g^{-1}(y))$$
$$= \sum_{x \in g^{-1}(y)} \mathsf{P}(X = x),$$

since there are only countably many non-zero contributions to this sum. Thus, if $Y = aX + b$ where $a \neq 0$ then

$$P(Y = y) = P(aX + b = y) = P(X = a^{-1}(y - b)) \text{ for all } y,$$

while if $Y = X^2$ then

$$P(Y = y) = \begin{cases} P(X = \sqrt{y}) + P(X = -\sqrt{y}) & \text{if } y > 0, \\ P(X = 0) & \text{if } y = 0, \\ 0 & \text{if } y < 0. \end{cases}$$

Exercises 8. Let X be a discrete random variable having the Poisson distribution with parameter λ, and let $Y = |\sin(\frac{1}{2}\pi X)|$. Find the mass function of Y.

2.4 Expectation

Consider a fair die. If it were thrown a large number of times, each of the possible outcomes $1, 2, \ldots, 6$ would appear on about one sixth of the throws and the average of the numbers observed would be approximately

$$\tfrac{1}{6} \cdot 1 + \tfrac{1}{6} \cdot 2 + \cdots + \tfrac{1}{6} \cdot 6 = 3\tfrac{1}{2},$$

which we call the *mean value*. This notion of mean value is easily extended to more general distributions as follows.

If X is a discrete random variable, the *expectation* of X is denoted by $E(X)$ and defined by

(19)
$$E(X) = \sum_{x \in \mathrm{Im}\, X} x P(X = x)$$

whenever this sum converges absolutely (in that $\sum_x |x P(X = x)| < \infty$).
    Equation (19) is often written

$$E(X) = \sum_x x P(X = x) = \sum_x x p_X(x),$$

and the expectation of X is often called the *expected value* or *mean* of X. The reason for requiring absolute convergence in (19) is that the image $\mathrm{Im}\, X$ of X may be an infinite set, and we need the summation in (19) to take the same value irrespective of the order in which we add up its terms.
    The physical analogy of 'expectation' is the idea of 'centre of gravity'. If masses with weights $\pi_1, \pi_2, \ldots$ are placed at the points $x_1, x_2, \ldots$ of $\mathbb{R}$, then the position of the centre of gravity is $\sum \pi_i x_i / \sum \pi_i$, or $\sum x_i p_i$, where $p_i = \pi_i / \sum_j \pi_j$ is the proportion of the total weight allocated to position x_i.

If X is a discrete random variable (on some probability space) and $g : \mathbb{R} \to \mathbb{R}$, then $Y = g(X)$ is a discrete random variable also. According to the above definition, we need to know the mass function of Y before we can calculate its expectation. The following theorem provides a useful way of avoiding this tedious calculation.

Theorem 2B   *If X is a discrete random variable and $g : \mathbb{R} \to \mathbb{R}$ then*

$$E(g(X)) = \sum_{x \in \text{Im } X} g(x)P(X = x),$$

whenever this sum converges absolutely.

Intuitively this result is rather clear, since $g(X)$ takes the value $g(x)$ when X takes the value x, an event which has probability $P(X = x)$. A more formal proof proceeds as follows.

Proof   Writing I for the image of X, we have that $Y = g(X)$ has image $g(I)$. Thus

$$E(Y) = \sum_{y \in g(I)} yP(Y = y)$$

$$= \sum_{y \in g(I)} y \sum_{\substack{x \in I: \\ g(x) = y}} P(X = x) \quad \text{by (18)}$$

$$= \sum_{x \in I} g(x)P(X = x)$$

if the last sum converges absolutely. $\qquad\square$

Here is an example of this theorem in action.

Example 20   Suppose that X is a random variable with the Poisson distribution, parameter λ, and we wish to find the expected value of $Y = e^X$. Without Theorem 2B we should have to find the mass function of Y. Actually this is not difficult, but it is even easier to apply the theorem to find that

$$E(Y) = E(e^X)$$

$$= \sum_{k=0}^{\infty} e^k P(X = k)$$

$$= \sum_{k=0}^{\infty} e^k \frac{1}{k!} \lambda^k e^{-\lambda}$$

$$= e^{-\lambda} \sum_{k=0}^{\infty} \frac{1}{k!} (\lambda e)^k$$

$$= e^{\lambda(e-1)}. \qquad\square$$

The expectation $E(X)$ of a discrete random variable X is an indication of the 'centre' of the distribution of X. Another important quantity associated with X is the 'variance' of X, and this is a measure of the degree of dispersion of X about its expectation $E(X)$. Formally, the *variance* of a discrete random variable X is defined to be the expectation of $[X - E(X)]^2$; the variance of X is usually written as

(21)
$$\text{var}(X) = E([X - E(X)]^2)$$

and we note that, by Theorem 2B,

(22)
$$\text{var}(X) = \sum_{x \in \text{Im} X} (x - \mu)^2 P(X = x)$$

where

$$\mu = E(X) = \sum_{x \in \text{Im} X} x P(X = x).$$

A rough motivation for this definition is as follows. If the dispersion of X about its expectation is very small then $|X - \mu|$ tends to be small, giving that $\text{var}(X) = E(|X - \mu|^2)$ is small also; on the other hand if there is often a considerable difference between X and its mean then $|X - \mu|$ may be large, giving that $\text{var}(X)$ is large also.

Equation (22) is not always the most convenient way to calculate the variance of a discrete random variable. We may expand the term $(x - \mu)^2$ in (22) to obtain

$$\text{var}(X) = \sum_x (x^2 - 2\mu x + \mu^2) P(X = x)$$

$$= \sum_x x^2 P(X = x) - 2\mu \sum_x x P(X = x) + \mu^2 \sum_x P(X = x)$$

$$= E(X^2) - 2\mu^2 + \mu^2 \qquad \text{by (19) and (5)}$$

$$= E(X^2) - \mu^2,$$

where $\mu = E(X)$ as before. Hence we obtain the useful formula

(23)
$$\text{var}(X) = E(X^2) - E(X)^2.$$

Example 24 If X has the geometric distribution with parameter p ($=1 - q$) then the mean of X is†

$$E(X) = \sum_{k=1}^{\infty} kpq^{k-1}$$

$$= \frac{p}{(1-q)^2} = \frac{1}{p},$$

† To sum a series such as $\sum_{k=0}^{\infty} kx^{k-1}$, just note that if $|x| < 1$ then $\sum_k kx^{k-1} = (d/dx) \sum_k x^k$, and hence $\sum_{k=0}^{\infty} kx^{k-1} = (d/dx)(1-x)^{-1} = (1-x)^{-2}$. Repeated differentiation of $(1-x)^{-1}$ yields formulae for $\sum_k k(k-1)x^{k-2}$ and other summations.

and the variance of X is

$$\text{var}(X) = \sum_{k=1}^{\infty} k^2 pq^{k-1} - \frac{1}{p^2}$$

by (23). However,

$$\sum_{k=1}^{\infty} k^2 q^{k-1} = q \sum_{k=1}^{\infty} k(k-1)q^{k-2} + \sum_{k=1}^{\infty} kq^{k-1}$$

$$= \frac{2q}{(1-q)^3} + \frac{1}{(1-q)^2}$$

by the last footnote, giving that

$$\text{var}(X) = p\left(\frac{2q}{p^3} + \frac{1}{p^2}\right) - \frac{1}{p^2}$$

$$= qp^{-2}. \qquad \square$$

Exercises   9. If X has the binomial distribution with parameters n and $p = 1 - q$, show that

$$\text{E}(X) = np, \qquad \text{E}(X^2) = npq + n^2 p^2,$$

and deduce the variance of X.

10. Find $\text{E}(X)$ and $\text{E}(X^2)$ when X has the Poisson distribution with parameter λ, and hence show that the Poisson distribution has variance equal to its mean.

2.5   Conditional expectation and the partition theorem

Suppose that X is a discrete random variable on the probability space $(\Omega, \mathcal{F}, \text{P})$, and that B is an event with $\text{P}(B) > 0$. If we are given that B occurs then this information affects the probability distribution of X; that is, probabilities such as $\text{P}(X = x)$ are replaced by conditional probabilities such as $\text{P}(X = x \mid B) = \text{P}(\{\omega \in \Omega : X(\omega) = x\} \cap B)/\text{P}(B)$.

If X is a discrete random variable and $\text{P}(B) > 0$, then the *conditional expectation of X given B* is denoted by $\text{E}(X \mid B)$ and defined by

(25) $$\text{E}(X \mid B) = \sum_{x \in \text{Im} \, X} x \text{P}(X = x \mid B),$$

whenever this sum converges absolutely.

Just as Theorem 1B expressed probabilities in terms of conditional probabilities, so expectations may be expressed in terms of conditional expectations.

Theorem 2C *If X is a discrete random variable and $\{B_1, B_2, \ldots\}$ is a partition of the sample space such that $P(B_i) > 0$ for each i, then*

(26)
$$E(X) = \sum_i E(X \mid B_i)P(B_i),$$

whenever this sum converges absolutely.

Proof The right-hand side of (26) equals, by (25),

$$\sum_i \sum_x xP(\{X = x\} \cap B_i) = \sum_x xP\left(\{X = x\} \cap \left(\bigcup_i B_i\right)\right)$$

$$= \sum_x xP(X = x). \qquad \square$$

We close this chapter with an example of this partition theorem in use.

Example 27 A coin is tossed repeatedly, and heads appears at each toss with probability p where $0 < p = 1 - q < 1$. Find the expected length of the initial run (this is a run of heads if the first toss gives heads, and of tails otherwise).

Solution Let H be the event that the first toss gives heads and H^c the event that the first toss gives tails. The pair H, H^c forms a partition of the sample space. Let X be the length of the initial run. It is easy to see that

$$P(X = k \mid H) = p^{k-1}q \quad \text{for } k = 1, 2, \ldots,$$

since if H occurs then $X = k$ if and only if the first toss is followed by exactly $k - 1$ heads and then a tail. Similarly

$$P(X = k \mid H^c) = q^{k-1}p \quad \text{for } k = 1, 2, \ldots.$$

Thus

$$E(X \mid H) = \sum_{k=1}^{\infty} kp^{k-1}q = \frac{q}{(1-p)^2} = \frac{1}{q},$$

and similarly

$$E(X \mid H^c) = \frac{1}{p}.$$

By Theorem 2C

$$E(X) = E(X \mid H)P(H) + E(X \mid H^c)P(H^c)$$

$$= \frac{1}{q}p + \frac{1}{p}q$$

$$= \frac{1}{pq} - 2. \qquad \square$$

Exercises 11. Let X be a discrete random variable and let g be a function from $\mathbb{R}$ to $\mathbb{R}$. If x is a real number such that $P(X = x) > 0$, show formally that

$$E(g(X) \mid X = x) = g(x),$$

and deduce from the partition theorem (Theorem 2C) that

$$E(g(X)) = \sum_x g(x)P(X = x).$$

12. Let N be the number of tosses of a fair coin up to and including the appearance of the first head. By conditioning on the result of the first toss, show that $E(N) = 2$.

2.6 Problems

1. If X has the Poisson distribution with parameter λ, show that

$$E(X(X-1)(X-2) \cdots (X-k)) = \lambda^{k+1}$$

for $k = 0, 1, 2, \ldots$.
2. Each toss of a coin results in heads with probability p (>0). If $m(r)$ is the mean number of tosses up to and including the rth head, show that

$$m(r) = p(1 + m(r-1)) + (1-p)(1 + m(r))$$

for $r = 1, 2, \ldots$, with the convention that $m(0) = 0$. Solve this difference equation by the method described in the appendix.
3. If X is a discrete random variable and $E(X^2) = 0$, show that $P(X = 0) = 1$. Deduce that if $\text{var}(X) = 0$ then $P(X = \mu) = 1$, where $\mu = E(X)$.
4. For what values of c and α is the function p, defined by

$$p(k) = \begin{cases} ck^\alpha & \text{for } k = 1, 2, \ldots, \\ 0 & \text{otherwise}, \end{cases}$$

a mass function?
5. If X has the geometric distribution with parameter p, show that

$$P(X > m + n \mid X > m) = P(X > n)$$

for $m, n = 0, 1, 2, \ldots$. We say that X has the 'lack-of-memory property' since, if we are given that $X - m > 0$, then the distribution of $X - m$ is the same as the original distribution of X. Show that the geometric distribution is the only distribution concentrated on the positive integers with the lack-of-memory property.
6. The random variable N takes non-negative integer values. Show that

$$E(N) = \sum_{k=0}^{\infty} P(N > k)$$

provided that the series on the right-hand side converges.
A fair die having two faces coloured blue, two red, and two green, is thrown repeatedly. Find the probability that not all colours occur in the first k throws.
Deduce that, if N is the random variable which takes the value n if all

three colours occur in the first n throws but only two of the colours in the first $n-1$ throws, then the expected value of N is $\frac{11}{2}$.   (Oxford 1979M)

7. *Coupon-collecting problem.* There are c different types of coupon and each time you obtain a coupon it is equally likely to be any one of the c types. Find the probability that the first n coupons which you collect do not form a complete set, and·deduce an expression for the mean number of coupons you will need to collect before you have a complete set.

*8. An ambidextrous student has a left and a right pocket, each initially containing n humbugs. Each time he feels hungry he puts a hand into one of his pockets and if it is not empty, takes a humbug from it and eats it. On each occasion, he is equally likely to choose either the left or right pocket. When he first puts his hand into an empty pocket the other pocket contains H humbugs.
Show that if p_h is the probability that $H = h$, then

$$p_h = \binom{2n-h}{n}\frac{1}{2^{2n-h}},$$

and find the expected value of H, by considering

$$\sum_{h=0}^{n} p_h, \quad \sum_{h=0}^{n} hp_h, \quad \sum_{h=0}^{n} (n-h)p_h$$

or otherwise.   (Oxford 1982M)

9. The probability of obtaining a head when a certain coin is tossed is p. The coin is tossed repeatedly until n heads occur in a row. Let X be the total number of tosses required for this to happen. Find the expected value of X.

10. A population of N animals has had a certain number a of its members captured, marked, and then released. Show that the probability P_n that it is necessary to capture n animals in order to obtain m which have been marked is

$$P_n = \frac{a}{N}\binom{a-1}{m-1}\binom{N-a}{n-m}\Big/\binom{N-1}{n-1},$$

where $m \le n \le N - a + m$. Hence show that

$$\frac{a}{N}\binom{a-1}{m-1}\frac{(N-a)!}{(N-1)!}\sum_{n=m}^{N-a+m}\frac{(n-1)!\,(N-n)!}{(n-m)!\,(N-a+m-n)!} = 1,$$

and that the expectation of n is $\dfrac{N+1}{a+1}m$.   (Oxford 1972M)

3

Multivariate discrete distributions and independence

3.1 Bivariate discrete distributions

Let X and Y be discrete random variables on the probability space $(\Omega, \mathscr{F}, \mathsf{P})$. Instead of treating X and Y separately, it is often necessary to regard the pair (X, Y) as a random vector taking values in $\mathbb{R}^2$.

If X and Y are discrete random variables on $(\Omega, \mathscr{F}, \mathsf{P})$, the *joint (probability) mass function* $p_{X,Y}$ of X and Y is the function $p_{X,Y}: \mathbb{R}^2 \to [0, 1]$ defined by

(1)
$$p_{X,Y}(x, y) = \mathsf{P}(\{\omega \in \Omega : X(\omega) = x \text{ and } Y(\omega) = y\}).$$

Thus $p_{X,Y}(x, y)$ is the probability that $X = x$ and $Y = y$, and we often write this as

$$p_{X,Y}(x, y) = \mathsf{P}(X = x, Y = y).$$

If $p_{X,Y}$ is the joint mass function of X and Y then it is clear that

(2)
$$p_{X,Y}(x, y) = 0 \quad \text{unless} \quad x \in \operatorname{Im} X \quad \text{and} \quad y \in \operatorname{Im} Y,$$

(3)
$$\sum_{x \in \operatorname{Im} X} \sum_{y \in \operatorname{Im} Y} p_{X,Y}(x, y) = 1.$$

The individual mass functions p_X and p_Y of X and Y may be found from $p_{X,Y}$ thus:

(4)
$$p_X(x) = \mathsf{P}(X = x) = \sum_{y \in \operatorname{Im} Y} \mathsf{P}(X = x, Y = y)$$

$$= \sum_y p_{X,Y}(x, y),$$

and similarly

(5)
$$p_Y(y) = \sum_x p_{X,Y}(x, y).$$

These mass functions, given by (4) and (5), are called the *marginal* mass functions of X and Y respectively since, if we think of (X, Y) as a randomly chosen point in the plane, then X and Y are the projections of this point onto the co-ordinate axes.

Example 6   Suppose that X and Y are random variables each taking the values 1, 2, or 3, and that the probability that the pair (X, Y) equals (x, y) is given by the following table for all relevant values of x and y.

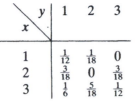

	y 1	2	3
x			
1	$\frac{1}{12}$	$\frac{1}{18}$	0
2	$\frac{3}{18}$	0	$\frac{3}{18}$
3	$\frac{1}{6}$	$\frac{5}{18}$	$\frac{1}{12}$

Then, for example,

$$P(X=3) = P(X=3,\ Y=1) + P(X=3,\ Y=2) + P(X=3,\ Y=3)$$
$$= \tfrac{1}{6} + \tfrac{5}{18} + \tfrac{1}{12} = \tfrac{19}{36}.$$

Similarly

$$P(Y=2) = \tfrac{1}{18} + 0 + \tfrac{5}{18} = \tfrac{1}{3}. \qquad\qquad \square$$

Similar ideas apply to families $X = (X_1, X_2, \ldots, X_n)$ of discrete random variables on a probability space. For example, the *joint mass function* of X is a function p_X defined by

$$p_X(x) = P(X_1 = x_1,\ X_2 = x_2,\ \ldots,\ X_n = x_n)$$

for $x = (x_1, x_2, \ldots, x_n) \in \mathbb{R}^n$.

Exercises   1. Two cards are drawn at random from a deck of 52 cards. If X denotes the number of aces drawn and Y denotes the number of kings, display the joint mass function of X and Y in the tabular form of Example 6.
2. The pair of discrete random variables (X, Y) have joint mass function

$$P(X = i,\ Y = j) = \begin{cases} \theta^{i+j+1} & \text{if } i, j = 0, 1, 2, \\ 0 & \text{otherwise,} \end{cases}$$

for some value of θ. Show that θ satisfies the equation

$$\theta + 2\theta^2 + 3\theta^3 + 2\theta^4 + \theta^5 = 1,$$

and find the marginal mass function of X in terms of θ.

3.2   Expectation in the multivariate case

If X and Y are discrete random variables on $(\Omega, \mathscr{F}, P)$ and $g : \mathbb{R}^2 \to \mathbb{R}$ then it is easy to check that $Z = g(X, Y)$ is a discrete random variable on $(\Omega, \mathscr{F}, P)$ also, defined formally by $Z(\omega) = g(X(\omega), Y(\omega))$ for $\omega \in \Omega$. The expectation of Z may be calculated directly from the joint mass function $p_{X,Y}(x, y) = P(X = x, Y = y)$ of X and Y, as the following theorem indicates; the proof is exactly analogous to that of Theorem 2B.

Theorem 3A
$$E(g(X, Y)) = \sum_{x \in \mathrm{Im}X} \sum_{y \in \mathrm{Im}Y} g(x, y) P(X = x, Y = y),$$

whenever this sum converges absolutely.

One particular consequence of this is of great importance: the expectation operator E acts linearly on the set of discrete random variables. That is to say, if X and Y are discrete random variables on $(\Omega, \mathscr{F}, P)$, and $a, b \in \mathbb{R}$, then

(7)
$$E(aX + bY) = aE(X) + bE(Y),$$

whenever $E(X)$ and $E(Y)$ exist. To see this, we use Theorem 3A with $g(x, y) = ax + by$ to find that

$$E(aX + bY) = \sum_x \sum_y (ax + by) P(X = x, Y = y)$$

$$= a \sum_x x \sum_y P(X = x, Y = y) + b \sum_y y \sum_x P(X = x, Y = y)$$

$$= a \sum_x xP(X = x) + b \sum_y yP(Y = y) \quad \text{by (4) and (5)}$$

$$= aE(X) + bE(Y).$$

Exercises   3. Suppose that (X, Y) has joint mass function

$$P(X = i, Y = j) = \theta^{i+j+1} \quad \text{for } i, j = 0, 1, 2.$$

Show that

$$E(XY) = \theta^3 + 4\theta^4 + 4\theta^5$$

and

$$E(X) = \theta^2 + 3\theta^3 + 3\theta^4 + 2\theta^5.$$

3.3   Independence of discrete random variables

In a probability space $(\Omega, \mathscr{F}, P)$, events A and B are called independent if $P(A \cap B) = P(A)P(B)$. Discrete random variables X and Y on $(\Omega, \mathscr{F}, P)$ are called 'independent' if the value taken by X is independent of the value taken by Y. That is to say, X and Y are *independent* if the pair of events $\{\omega \in \Omega : X(\omega) = x\}$ and $\{\omega \in \Omega : Y(\omega) = y\}$ are independent for all $x, y \in \mathbb{R}$, and we normally write this condition as

(8)
$$P(X = x, Y = y) = P(X = x)P(Y = y) \quad \text{for all } x, y \in \mathbb{R}.$$

Random variables which are not independent are called *dependent*.

Condition (8) may be expressed as

(9) $$p_{X,Y}(x,y) = \left(\sum_y p_{X,Y}(x,y)\right)\left(\sum_x p_{X,Y}(x,y)\right) \qquad \text{for all } x, y \in \mathbb{R},$$

in terms of the joint mass function of X and Y; this latter condition may be simplified as indicated by the following theorem.

Theorem 3B *Discrete random variables X and Y are independent if and only if there exist functions $f, g : \mathbb{R} \to \mathbb{R}$ such that the joint mass function of X and Y satisfies*

(10) $$p_{X,Y}(x,y) = f(x)g(y) \qquad \text{for all } x, y \in \mathbb{R}.$$

Of course, we need only check (10) for $x \in \text{Im } X$ and $y \in \text{Im } Y$.

Proof We need only prove the sufficiency of the condition. Suppose that (10) holds for some f and g. By (4) and (5)

$$p_X(x) = f(x)\sum_y g(y), \qquad p_Y(y) = g(y)\sum_x f(x),$$

and by (3)

$$1 = \sum_{x,y} p_{X,Y}(x,y) = \sum_{x,y} f(x)g(y)$$
$$= \sum_x f(x) \sum_y g(y).$$

Therefore

$$p_{X,Y}(x,y) = f(x)g(y)$$
$$= f(x)g(y)\sum_x f(x)\sum_y g(y)$$
$$= p_X(x)p_Y(y). \qquad \square$$

Example 11 Suppose that X and Y are random variables each taking values in the non-negative integers with joint mass function

$$p_{X,Y}(i,j) = P(X=i, Y=j) = \frac{1}{i!\,j!}\lambda^i\mu^j e^{-(\lambda+\mu)} \qquad \text{for } i,j = 0, 1, 2, \dots.$$

Immediate from Theorem 3B is the fact that X and Y are independent, since their joint mass function may be factorized in the form

$$p_{X,Y}(i,j) = \left(\frac{1}{i!}\lambda^i\right)\left(\frac{1}{j!}\mu^j e^{-(\lambda+\mu)}\right),$$

as a function of i multiplied by a function of j. Such a factorization is not unique of course, and it is more natural to write

$$p_{X,Y}(i,j) = \left(\frac{1}{i!}\lambda^i e^{-\lambda}\right)\left(\frac{1}{j!}\mu^j e^{-\mu}\right)$$

as the product of the marginal mass functions: X and Y are independent random variables each having a Poisson distribution, with parameters λ and μ respectively. $\square$

The following is an important property of independent pairs of random variables.

Theorem 3C *If X and Y are independent discrete random variables with expectations $E(X)$ and $E(Y)$ then*

$$E(XY) = E(X)E(Y).$$

Proof By Theorem 3A,

$$E(XY) = \sum_{x,y} xy P(X=x, Y=y)$$

$$= \sum_{x,y} xy P(X=x)P(Y=y) \qquad \text{by independence}$$

$$= \sum_{x} x P(X=x) \sum_{y} y P(Y=y)$$

$$= E(X)E(Y).$$

It is the existence of $E(X)$ and $E(Y)$ which authorizes us to interchange the summations as we have done. $\square$

The converse of Theorem 3C is false: if $E(XY) = E(X)E(Y)$ then it does not follow that X and Y are independent (see Example 13 below). The correct converse is given next.

Theorem 3D *Discrete random variables X and Y on $(\Omega, \mathcal{F}, P)$ are independent if and only if*

(12) $$E(g(X)h(Y)) = E(g(X))E(h(Y))$$

for all functions $g, h: \mathbb{R} \to \mathbb{R}$ for which the latter two expectations exist.

Proof The necessity of (12) follows just as in the proof of Theorem 3C. To prove sufficiency, choose $a, b \in \mathbb{R}$ and define g and h by

$$g(x) = \begin{cases} 1 & \text{if } x = a, \\ 0 & \text{if } x \neq a, \end{cases}$$

$$h(y) = \begin{cases} 1 & \text{if } y = b, \\ 0 & \text{if } y \neq b. \end{cases}$$

Then
$$E(g(X)h(Y)) = P(X = a, Y = b)$$
and
$$E(g(X))E(h(Y)) = P(X = a)P(Y = b)$$
giving by (12) that
$$p_{X,Y}(a, b) = p_X(a)p_Y(b). \qquad \square$$

Here is an example of two discrete random variables X and Y which are not independent but which satisfy $E(XY) = E(X)E(Y)$.

Example 13

Suppose that X has distribution given by
$$P(X = -1) = P(X = 0) = P(X = 1) = \tfrac{1}{3}$$
and Y is given by
$$Y = \begin{cases} 0 & \text{if } X = 0, \\ 1 & \text{if } X \neq 0. \end{cases}$$

It is easy to find a probability space $(\Omega, \mathcal{F}, P)$ together with two random variables having these distributions. For example, take $\Omega = \{-1, 0, 1\}$, $\mathcal{F}$ the set of all subsets of Ω, P given by $P(-1) = P(0) = P(1) = \tfrac{1}{3}$, and $X(\omega) = \omega$, $Y(\omega) = |\omega|$. Then X and Y are dependent since
$$P(X = 0, Y = 1) = 0$$
but
$$P(X = 0)P(Y = 1) = \tfrac{1}{3} \cdot \tfrac{2}{3} = \tfrac{2}{9}.$$

On the other hand
$$E(XY) = \sum_{x,y} xy P(X = x, Y = y)$$
$$= (-1) \cdot \tfrac{1}{3} + 0 \cdot \tfrac{1}{3} + 1 \cdot \tfrac{1}{3} = 0$$
and
$$E(X)E(Y) = 0 \cdot \tfrac{2}{3} = 0. \qquad \square$$

In this section we have considered pairs of random variables only, but the same ideas apply to vectors $X = (X_1, X_2, \ldots, X_n)$ of random variables where $n > 2$. For example, the family X is called *independent* if
$$P(X_1 = x_1, \ldots, X_n = x_n) = P(X_1 = x_1) \cdots P(X_n = x_n)$$
(or equivalently $p_X(x) = \prod_{i=1}^{n} p_{X_i}(x_i)$) for all $x = (x_1, \ldots, x_n) \in \mathbb{R}^n$. Furthermore, if $X_1, X_2, \ldots, X_n$ are independent then
$$E(X_1 X_2 \cdots X_n) = E(X_1)E(X_2) \cdots E(X_n),$$
just as in Theorem 3C. Finally, the family is called *pairwise independent* if X_i and X_j are independent whenever $i \neq j$.

Exercises 4. Let X and Y be independent discrete random variables. Prove that

$$P(X \geq x \text{ and } Y \geq y) = P(X \geq x)P(Y \geq y)$$

for all $x, y \in \mathbb{R}$.

5. The *indicator function* of an event A is the function χ_A defined by

$$\chi_A(\omega) = \begin{cases} 1 & \text{if } \omega \in A, \\ 0 & \text{if } \omega \notin A. \end{cases}$$

Show that two events A and B are independent if and only if their indicator functions are independent random variables.

6. If X and Y are independent discrete random variables, show that the two random variables $g(X)$ and $h(Y)$ are independent also, for any functions g and h which map $\mathbb{R}$ into $\mathbb{R}$.

3.4 Sums of random variables

Much of probability theory is concerned with sums of random variables, and so we need an answer to the following question: if X and Y are discrete random variables with a certain joint mass function, what is the mass function of $Z = X + Y$? Clearly Z takes the value z if and only if $X = x$ and $Y = z - x$ for some value of x, and so

(14)
$$P(Z = z) = P\left(\bigcup_x (\{X = x\} \cap \{Y = z - x\})\right)$$

$$= \sum_{x \in \text{Im}X} P(X = x, Y = z - x) \qquad \text{for } z \in \mathbb{R}.$$

If X and Y are independent then their joint mass function factorizes, and we obtain the following result.

Theorem 3E *If X and Y are independent discrete random variables on $(\Omega, \mathscr{F}, P)$ then $Z = X + Y$ has mass function*

(15)
$$P(Z = z) = \sum_{x \in \text{Im}X} P(X = x)P(Y = z - x) \quad \text{for } z \in \mathbb{R}.$$

Formula (15) is rather inconvenient in practice, since it involves a summation. Soon we shall see a better way of treating sums of independent random variables. In the language of analysis, formula (15) says that the mass function of $X + Y$ is the *convolution* of the mass functions of X and of Y.

Exercises 7. If X and Y are independent discrete random variables, X having the Poisson distribution with parameter λ and Y having the Poisson distribution with parameter μ, show that $X + Y$ has the Poisson distribution with parameter $\lambda + \mu$. Give an example to show that the conclusion is not generally true if X and Y are dependent.

8. If X has the binomial distribution with parameters m and p, Y has the binomial distribution with parameters n and p, and X and Y are independent, show that $X + Y$ has the binomial distribution with parameters $m + n$ and p.

9. Show by induction that the sum of n independent random variables, each having the Bernoulli distribution with parameter p, has the binomial distribution with parameters n and p.

3.5 Problems

1. Let X and Y be independent discrete random variables each having mass function given by

$$P(X = k) = P(Y = k) = pq^k \qquad \text{for } k = 0, 1, 2, \ldots,$$

where $0 < p = 1 - q < 1$. Show that

$$P(X = k \mid X + Y = n) = \frac{1}{n+1} \qquad \text{for } k = 0, 1, 2, \ldots, n.$$

2. Independent random variables U and V each take the values -1 or 1 only, and

$$P(U = 1) = a, \qquad P(V = 1) = b,$$

where $0 < a, b < 1$. A third random variable W is defined by $W = UV$. Show that there are unique values of a and b such that U, V, and W are pairwise independent. For these values of a and b, are U, V, and W independent? Justify your answer. (Oxford 1971F)

3. If X and Y are discrete random variables, each taking only two distinct values, prove that X and Y are independent if and only if $E(XY) = E(X)E(Y)$.

4. Let $X_1, X_2, \ldots, X_n$ be independent discrete random variables, each having mass function

$$P(X_i = k) = 1/N \qquad \text{for } k = 1, 2, \ldots, N.$$

Find the mass functions of U_n and V_n, given by

$$U_n = \min\{X_1, X_2, \ldots, X_n\}, \qquad V_n = \max\{X_1, X_2, \ldots, X_n\}.$$

5. Let X and Y be independent discrete random variables, X having the geometric distribution with parameter p and Y having the geometric distribution with parameter r. Show that $U = \min\{X, Y\}$ has the geometric distribution with parameter $p + r - pr$.

6. Let $X_1, X_2, \ldots$ be discrete random variables, each having mean μ, and let N be a random variable which takes values in the non-negative integers and which is independent of the X's. By conditioning on the value of N, show that

$$E(X_1 + X_2 + \cdots + X_N) = \mu E(N).$$

7. The random variables U and V each take the values ± 1. Their joint distribution is given by

$$P(U = +1) = P(U = -1) = \tfrac{1}{2},$$
$$P(V = +1 \mid U = 1) = \tfrac{1}{3} = P(V = -1 \mid U = -1),$$
$$P(V = -1 \mid U = 1) = \tfrac{2}{3} = P(V = +1 \mid U = -1).$$

   (a) Find the probability that $x^2 + Ux + V = 0$ has at least one real root.
   (b) Find the expected value of the larger root given that there is at least one real root.
   (c) Find the probability that $x^2 + (U + V)x + U + V = 0$ has at least one real root. (Oxford 1980M)

8. N balls are thrown at random into M boxes, with multiple occupancy permitted. Show that the expected number of empty boxes is $(M - 1)^N/M^{N-1}$.

9. We are provided with a coin which comes up heads with probability p at each toss. Let $v_1, v_2, \ldots, v_n$ be n distinct points on a unit circle. We examine each unordered pair v_i, v_j in turn and toss the coin; if it comes up heads, we join v_i and v_j by a straight line segment (called an *edge*) otherwise we do nothing. The resulting network is called a *random graph*. Prove that
   (i) the expected number of edges in the random graph is $\tfrac{1}{2}n(n - 1)p$,
   (ii) the expected number of triangles (triples of points each pair of which is joined by an edge) is $\tfrac{1}{6}n(n - 1)(n - 2)p^3$.

10. *Coupon-collecting problem.* There are c different types of coupon and each time you obtain a coupon it is equally likely to be any one of the c types. Let Y_i be the additional number of coupons collected, after obtaining i distinct types, before a new type is collected. Show that Y_i has the geometric distribution with parameter $(c - i)/c$, and deduce the mean number of coupons you will need to collect before you have a complete set.

11. In Problem 10 above, find the expected number of different types of coupon in the first n coupons received.

12. Each time you flip a certain coin, heads appears with probability p. Suppose that you flip the coin a random number N of times, where N has the Poisson distribution with parameter λ and is independent of the outcomes of the flips. Find the distributions of the numbers X and Y of resulting heads and tails, respectively, and show that X and Y are independent.

13. Let $(Z_n : 1 \le n < \infty)$ be a sequence of independent identically distributed random variables with

$$P(Z_n = 0) = q, \qquad P(Z_n = 1) = 1 - q.$$

Let A_i be the event that $Z_i = 0$ and $Z_{i-1} = 1$ for $i \ge 2$. If U_n is the number of times A_i occurs for $2 \le i \le n$ prove that $E(U_n) = (n - 1)q(1 - q)$ and find the variance of U_n. (Oxford 1977F)

4

Probability generating functions

4.1 Generating functions

One way of remembering a sequence $u_0, u_1, u_2, \ldots$ of real numbers is to write down a general formula for the nth term u_n. Another is to write down the *generating function* of the sequence, defined to be the sum of the power series

(1)
$$u_0 + u_1 s + u_2 s^2 + \cdots$$

whenever s is such that this series converges. For example, the sequence $1, 2, 4, 8, \ldots$ has generating function

$$1 + 2s + 4s^2 + \cdots = \sum_{n=0}^{\infty} (2s)^n$$
$$= (1 - 2s)^{-1},$$

valid whenever $|s| < \frac{1}{2}$. Similarly the sequence $1, 2, 3, \ldots$ has generating function

(2)
$$1 + 2s + 3s^2 + \cdots = \sum_{n=0}^{\infty} (n + 1)s^n$$
$$= (1 - s)^{-2},$$

valid whenever $|s| < 1$. Such generating functions are useful ways of dealing with real sequences since they specify the sequence uniquely. That is to say, given the sequence $u_0, u_1, u_2, \ldots$, we may find its generating function (1); conversely if the generating function $U(s)$ has a convergent Taylor series

$$U(s) = u_0 + u_1 s + u_2 s^2 + \cdots$$

for all small s, then this expansion is unique and so $U(s)$ generates the sequence $u_0, u_1, u_2, \ldots$ only.

There are other types of generating functions also; for example, the *exponential generating function* of the real sequence $u_0, u_1, u_2, \ldots$ is defined to be the sum of the power series

$$u_0 + u_1 s + \frac{1}{2!} u_2 s^2 + \frac{1}{3!} u_3 s^3 + \cdots$$

whenever this series converges. We do not consider such generating functions in this chapter, but shall return to this point later.

When dealing with generating functions of real sequences, it is important that the underlying power series converges for some $s \neq 0$, but so long as this is the case we will not normally go to the length of saying for which values of s the power series is convergent. For example, we say that $(1-s)^{-2}$ is the generating function of the sequence $1, 2, 3, \ldots$ without explicit reference to the fact that the series in (2) converges only if $|s| < 1$.

Example 3   The sequence given by

$$u_n = \begin{cases} \binom{N}{n} & \text{if } n = 0, 1, 2, \ldots, N \\ 0 & \text{otherwise} \end{cases}$$

has generating function

$$U(s) = \sum_{n=0}^{N} \binom{N}{n} s^n = (1+s)^N. \qquad \square$$

Exercises   1. If $u_0, u_1, \ldots$ has generating function $U(s)$ and $v_0, v_1, \ldots$ has generating function $V(s)$, find $V(s)$ in terms of $U(s)$ when (i) $v_n = 2u_n$, (ii) $v_n = u_n + 1$, (iii) $v_n = nu_n$.

2. Of which sequence is $U(s) = (1 - 4pqs^2)^{-\frac{1}{2}}$ the generating function (where $0 < p = 1 - q < 1$)?

4.2 Integer-valued random variables

Many random variables of interest take values in the set of non-negative integers (all the examples in Section 2.2 are of this form). We may think of the mass function of such a random variable X as a sequence $p_0, p_1, p_2, \ldots$ of numbers, where

$$p_k = P(X = k) \quad \text{for} \quad k = 0, 1, 2, \ldots,$$

satisfying

(4)
$$p_k \geq 0 \quad \text{for all } k, \text{ and } \sum_{k=0}^{\infty} p_k = 1.$$

The *probability generating function* of X is the function $G_X(s)$ defined by

(5)
$$G_X(s) = p_0 + p_1 s + p_2 s^2 + \cdots,$$

for all values of s for which the right-hand side converges absolutely.

In other words, the probability generating function $G_X(s)$ of X is the generating function of the sequence $p_0, p_1, \ldots$.

From (4) and (5) we see that

(6) $$G_X(0) = p_0 \quad \text{and} \quad G_X(1) = 1,$$

and the result of Theorem 2B provides the useful representation

(7) $$G_X(s) = E(s^X)$$

whenever this expectation exists. It is immediate that $G_X(s)$ exists for all values of s satisfying $|s| \leq 1$, since in this case

(8) $$\sum_{k=0}^{\infty} |p_k s^k| \leq \sum_{k=0}^{\infty} p_k = 1.$$

Example 9 Let X be a random variable having the geometric distribution with parameter p. Then

$$P(X = k) = pq^{k-1} \quad \text{for} \quad k = 1, 2, 3, \ldots$$

where $p + q = 1$, and X has probability generating function

$$G_X(s) = \sum_{k=1}^{\infty} pq^{k-1} s^k$$

$$= ps \sum_{k=0}^{\infty} (qs)^k$$

$$= \frac{ps}{1 - qs} \quad \text{if} \quad |s| < q^{-1}. \qquad \square$$

A crucially important property of probability generating functions is the following uniqueness theorem.

Theorem 4A *Suppose that X and Y have probability generating functions G_X and G_Y, respectively. Then*

$$G_X(s) = G_Y(s) \quad \text{for all } s$$

if and only if

$$P(X = k) = P(Y = k) \quad \text{for } k = 0, 1, 2, \ldots.$$

In other words, integer-valued random variables have the same probability generating function if and only if they have the same mass function.

Proof We need only show that $G_X = G_Y$ implies that $P(X = k) = P(Y = k)$ for all k. By (8), G_X and G_Y have radii of convergence at least 1 and therefore they have unique power series expansions about the origin:

$$G_X(s) = \sum_{k=0}^{\infty} s^k P(X = k), \qquad G_Y(s) = \sum_{k=0}^{\infty} s^k P(Y = k).$$

If $G_X = G_Y$ then these two power series have identical coefficients. ☐

In Example 9 we saw that a random variable with the geometric distribution, parameter p, has probability generating function $ps(1 - qs)^{-1}$, where $p + q = 1$. Only by an appeal to the above theorem can we deduce the converse: if X has probability generating function $ps(1 - qs)^{-1}$ then X has the geometric distribution with parameter p.

Here is a list of some common probability generating functions.

Bernoulli distribution. If X has this distribution with parameter p then

(10) $$G_X(s) = q + ps$$

where $p + q = 1$.

Binomial distribution. If X has this distribution with parameters n and p then

(11) $$G_X(s) = \sum_{k=0}^{n} \binom{n}{k} p^k q^{n-k} s^k$$

$$= (q + ps)^n$$

where $p + q = 1$.

Poisson distribution. If X has this distribution with parameter λ then

(12) $$G_X(s) = \sum_{k=0}^{\infty} \frac{1}{k!} \lambda^k e^{-\lambda} s^k$$

$$= e^{\lambda(s-1)}.$$

Negative binomial distribution. If X has this distribution with parameters n and p then

(13) $$G_X(s) = \sum_{k=n}^{\infty} \binom{k-1}{n-1} p^n q^{k-n} s^k$$

$$= \left(\frac{ps}{1 - qs} \right)^n \quad \text{if} \quad |s| < q^{-1},$$

where $p + q = 1$. We have used the negative binomial expansion here, as in the discussion after (2.10).

There are two principal reasons why it is often more convenient to work with probability generating functions than with mass functions, and we discuss these in the next two sections.

Exercises   3. If X is a random variable with probability generating function $G_X(s)$ and k

is a positive integer, show that $Y = kX$ and $Z = X + k$ have probability generating functions

$$G_Y(s) = G_X(s^k), \qquad G_Z(s) = s^k G_X(s).$$

4. If X is uniformly distributed on $\{0, 1, 2, \ldots, a\}$, so that

$$P(X = k) = \frac{1}{a+1} \quad \text{for } k = 0, 1, 2, \ldots, a,$$

show that X has probability generating function

$$G_X(s) = \frac{1 - s^{a+1}}{(a+1)(1-s)}.$$

4.3   Moments

For any discrete random variable X, the mean value $E(X)$ is an indication of the 'centre' of the distribution of X. This is only the first of a collection of numbers containing information about the distribution of X, the whole collection being the sequence $E(X)$, $E(X^2)$, $E(X^3)$, ... of means of powers of X. These numbers are called the *moments* of X. More formally, we define the kth *moment of X* to be $E(X^k)$, for $k = 1, 2, \ldots$. Possibly the two most important quantities which arise from the moments of X are the mean $E(X)$ of X, and the variance of X, defined in (2.21) to be

(14)
$$\text{var}(X) = E([X - E(X)]^2).$$

To see the relationship between $\text{var}(X)$ and the moments of X, just note that

(15)
$$\begin{aligned} \text{var}(X) &= E(X^2 - 2XE(X) + E(X)^2) \\ &= E(X^2) - 2E(X)^2 + E(X)^2 \qquad \text{by (3.7)} \\ &= E(X^2) - E(X)^2, \end{aligned}$$

in agreement with (2.23).

If X is a random variable taking values in the non-negative integers, the moments of X are easily found from the probability generating function of X by calculating the derivatives of this function at the point $s = 1$. The basic observation is as follows.

Theorem 4B   *Let X be a random variable with probability generating function $G_X(s)$. Then the rth derivative of $G_X(s)$ at $s = 1$ equals $E(X(X-1) \cdots (X-r+1))$ for $r = 1, 2, \ldots$; that is to say*

(16)
$$G_X^{(r)}(1) = E(X(X-1) \cdots (X-r+1)).$$

Proof To see this for the case when $r = 1$, we use the following unrigorous argument:

(17)
$$G'_X(s) = \frac{d}{ds} \sum_{k=0}^{\infty} s^k P(X = k)$$

$$= \sum_{k=0}^{\infty} \frac{d}{ds} s^k P(X = k)$$

$$= \sum_{k=0}^{\infty} ks^{k-1} P(X = k)$$

so that

$$G'_X(1) = \sum_{k=0}^{\infty} k P(X = k) = E(X)$$

as required. A similar argument holds for the rth derivative of $G_X(s)$ at $s = 1$. The difficulty in (17) is to justify the interchange of the differential operator and the summation, but this may be shown to be valid if $|s| < 1$, and then Abel's Lemma† enables us to conclude that (17) is all right. □

It is easy to see how to calculate the moments of X from (16). For example

(18)
$$E(X) = G'_X(1),$$

(19)
$$E(X^2) = E(X(X - 1) + X)$$
$$= E(X(X - 1)) + E(X)$$
$$= G''_X(1) + G'_X(1),$$

and also

(20)
$$\text{var}(X) = G''_X(1) + G'_X(1) - G'_X(1)^2$$

from (15).

Example 21 If X has the geometric distribution with parameter p then $G_X(s) = ps(1 - qs)^{-1}$ if $|s| < q^{-1}$, where $p + q = 1$. Hence

$$E(X) = G'_X(1) = p^{-1},$$
$$E(X^2) = G''_X(1) + G'_X(1) = (q + 1)p^{-2},$$
$$\text{var}(X) = qp^{-2},$$

in agreement with the calculations of Example 2.24. □

† Abel's Lemma is a classical result of real analysis. It says that if $u_0, u_1, \ldots$ is a real non-negative sequence such that the power series $\sum_{k=0}^{\infty} u_k s^k$ converges with sum $U(s)$ if $|s| < 1$, then $\sum_{k=0}^{\infty} u_k = \lim_{s \uparrow 1} U(s)$, where we allow the possibility that both sides equal $+\infty$.

Exercises   5. Use the method of generating functions to show that a random variable
having the Poisson distribution, parameter λ, has both mean and variance
equal to λ.

6. If X has the negative binomial distribution with parameters n and p, show
that
$$E(X) = n/p, \qquad \text{var}(X) = nq/p^2,$$
where $q = 1 - p$.

4.4 Sums of independent random variables

Much of probability theory is concerned with sums of independent
random variables, and we need a way of dealing with such sums. The
convolution argument of Theorem 3E is usually inconvenient, since
$n - 1$ convolutions are required to find the mass function of the sum
of n independent random variables and each such operation can be
rather complicated. It is in this respect that probability generating
functions are a very powerful tool.

Theorem   *If X and Y are independent random variables each taking values in*
4C $\{0, 1, 2, \ldots\}$, *then their sum has probability generating function*

(22)
$$G_{X+Y}(s) = G_X(s)G_Y(s).$$

Proof
$$\begin{aligned} G_{X+Y}(s) &= E(s^{X+Y}) \\ &= E(s^X s^Y) \\ &= E(s^X)E(s^Y) \qquad \text{by Theorem 3D} \\ &= G_X(s)G_Y(s). \end{aligned} \qquad \square$$

It follows that the sum $S_n = X_1 + \cdots + X_n$ of n independent
random variables, each taking values in $\{0, 1, 2, \ldots\}$, has probability
generating function given by

(23)
$$G_{S_n}(s) = G_{X_1}(s)G_{X_2}(s) \cdots G_{X_n}(s);$$

we shall make much use of this formula. An extension of (23) deals
with the sum of a random number of independent random variables.

Theorem   *Let $N, X_1, X_2, \ldots$ be independent random variables, each taking*
4D   *values in $\{0, 1, 2, \ldots\}$. If the X's are identically distributed, each*
having probability generating function G_X, then the sum
$$S = X_1 + X_2 + \cdots + X_N$$
has probability generating function

(24)
$$G_S(s) = G_N(G_X(s)).$$

Proof

We shall use the partition theorem 2C with the events $B_n = \{N = n\}$ for $n = 0, 1, 2, \ldots$. Thus

$$G_S(s) = E(s^{X_1 + \cdots + X_N})$$

$$= \sum_{n=0}^{\infty} E(s^{X_1 + \cdots + X_N} \mid N = n)\, P(N = n) \qquad \text{by Theorem 2C}$$

$$= \sum_{n=0}^{\infty} E(s^{X_1 + \cdots + X_n})\, P(N = n)$$

$$= \sum_{n=0}^{\infty} G_X(s)^n P(N = n) \qquad \text{by (23)}$$

$$= G_N(G_X(s))$$

by the definition of G_N.   □

Formula (24) enables us to say quite a lot about the sum of a random number of independent identically distributed random variables. For example, to find the mean value of S, in the notation of Theorem 4D, we merely calculate $G_S'(1)$ as follows. By (24)

$$G_S'(s) = G_N'(G_X(s))G_X'(s);$$

now set $s = 1$ to obtain

$$G_S'(1) = G_N'(G_X(1))G_X'(1)$$
$$= G_N'(1)G_X'(1),$$

since $G_X(1) = 1$, to obtain from (18) that

(25) $$E(S) = E(N)E(X)$$

where $E(X)$ is the mean of a typical one of the X's.

Example 26

One evening the hutch in the garden contains 20 pregnant rabbits. The hutch is insecure and each rabbit has a chance of $\frac{1}{2}$ of escaping overnight. The next morning each remaining rabbit gives birth to a litter, each mother having a random number of offspring with the Poisson distribution, parameter 3 (this is a very unlikely tale). Assuming as much independence as necessary, determine the probability generating function of the total number of offspring.

Solution

The number N of rabbits who do not escape has the binomial distribution with parameters 20 and $\frac{1}{2}$, and consequently N has probability generating function

$$G_N(s) = E(s^N) = (\tfrac{1}{2} + \tfrac{1}{2}s)^{20}.$$

Let X_i be the number of offspring of the ith of these rabbits; each X_i

has the Poisson distribution with probability generating function

$$G_X(s) = e^{3(s-1)}.$$

Assuming that N and the X's are independent, we conclude from the random sum formula (24) that the total number $S = X_1 + X_2 + \cdots + X_N$ of offspring has probability generating function

$$G_S(s) = G_N(G_X(s))$$
$$= (\tfrac{1}{2} + \tfrac{1}{2}e^{3(s-1)})^{20}. \qquad \square$$

Exercises
7. Use Theorem 4C to show that the sum of two independent random variables, having the Poisson distribution with parameters λ and μ respectively, has the Poisson distribution also, with parameter $\lambda + \mu$. Compare your solution to that of Exercise 7 at the end of Section 3.4.
8. Use generating functions to find the distribution of $X + Y$ where X and Y are independent random variables, X having the binomial distribution with parameters m and p, and Y having the binomial distribution with parameters n and p. Deduce that the sum of n independent random variables, each having the Bernoulli distribution with parameter p, has the binomial distribution with parameters n and p.
9. Each egg laid by the hen falls onto the concrete floor of the henhouse and cracks with probability p. If the number of eggs laid today by the hen has the Poisson distribution, parameter λ, use generating functions to show that the number of uncracked eggs has the Poisson distribution with parameter $\lambda(1 - p)$.

4.5 Problems

1. Let X have probability generating function $G_X(s)$ and let $u_n = P(X > n)$. Show that the generating function $U(s)$ of the sequence $u_0, u_1, \ldots$ satisfies

$$(1 - s)U(s) = 1 - G_X(s),$$

   whenever the series defining these generating functions converge.
2. A symmetrical die is thrown independently seven times; what is the probability that the total number of points obtained is 14?
   (Oxford 1974M)
3. Three players Alan, Bob, and Cindy throw a perfect die in turn independently in the order A, B, C, A, etc. until one wins by throwing a 5 or a 6. Show that the probability generating function $F(s)$ for the random variable X which takes the value r if the game ends on the rth throw can be written as

$$F(s) = \frac{9s}{27 - 8s^3} + \frac{6s^2}{27 - 8s^3} + \frac{4s^3}{27 - 8s^3}.$$

   Hence find the probabilities of winning for Alan, Bob, and Cindy. Find the mean length of the game.
   (Oxford 1973M)

4. A player undertakes trials and the probability of success at each trial is
   p. His turn consists of a sequence of trials up to the first failure. Obtain
   the probability generating function for the total number of successes in N
   turns. Show that the mean of this distribution is $Np(1-p)^{-1}$ and find its
   variance.   (Oxford 1974M)

5. Each year a tree of a particular type flowers once and the probability
   that it has n flowers is $(1-p)p^n$, $n = 0, 1, 2, \ldots$, where $0 < p < 1$. Each
   flower has probability $\frac{1}{2}$ of producing a ripe fruit, independently of all
   other flowers. Find the probability that in a given year
   (a) the tree produces r ripe fruits,
   (b) the tree had n flowers if it produces r ripe fruits.   (Oxford 1982M)

6. An unfair coin is tossed n times, each outcome is independent of all the
   others, and on each toss a head is shown with probability p. The total
   number of heads shown is X. Use the probability generating function of
   X to find:
   (a) the mean and variance of X,
   (b) the probability that X is even,
   (c) the probability that X is divisible by 3.   (Oxford 1980M)

7. Let X and Y be independent random variables having Poisson distributions with parameters λ and μ respectively. Prove that $X + Y$ has a
   Poisson distribution and that $\text{var}(X + Y) = \text{var}(X) + \text{var}(Y)$. Find the
   conditional probability $\text{P}(X = k \mid X + Y = n)$ for $0 \le k \le n$, and hence
   show that the conditional expectation of X given that $X + Y = n$, that is,

$$E(X \mid X + Y = n) = \sum_{k=0}^{\infty} k\text{P}(X = k \mid X + Y = n),$$

   is $n\lambda/(\lambda + \mu)$.   (Oxford 1983M)

*8. A fair coin is tossed a random number N of times giving a total of X
   heads and Y tails. You showed in Problem 12 of Section 3.5 that X and
   Y are independent if N has the Poisson distribution. Use generating
   functions to show that the converse is valid too: if X and Y are
   independent and the generating function $G_N(s)$ of N is assumed to exist
   for values of s in a neighbourhood of $s = 1$, then N has the Poisson
   distribution.

9. Each packet of a certain breakfast cereal contains one token, coloured
   either red, blue, or green. The coloured tokens are distributed randomly
   among the packets, each colour being equally likely. Let X be the
   random variable which takes the value j when I find my first red token in
   the jth packet which I open. Obtain the probability generating function
   of X, and hence find its expectation.
   More generally, suppose that there are tokens of m different colours, all
   equally likely. Let Y be the random variable which takes the value j
   when I first obtain a full set, of at least one token of each colour, when I
   open my jth packet. Find the generating function of Y, and show that its
   expectation is $m\left(1 + \dfrac{1}{2} + \dfrac{1}{3} + \cdots + \dfrac{1}{m}\right)$.   (Oxford 1985M)

10. Define the mean value of a discrete random variable and the probability
   generating function ϕ. Show that the mean value is $\phi'(1)$. If $\phi(s)$ has
   the form $p(s)/q(s)$ show that the mean value is $(p'(1) - q'(1))/q(1)$.
   Two duellists A and B fire at each other in turn until one hits the other.

Each duellist has the same probability of obtaining a hit with each shot he fires, these probabilities being a for A and b for B. If A fires the first shot calculate the probability that he wins the duel. Find also the probability distribution of the number of shots fired before the duel terminates. What is the expected number of shots fired?

(Oxford 1976M)

5

Distribution functions and density functions

5.1 Distribution functions

Discrete random variables may take only countably many values. This condition is too restrictive for many situations, and accordingly we make a broader definition: a *random variable* X on the probability space $(\Omega, \mathscr{F}, \mathsf{P})$ is a mapping $X : \Omega \to \mathbb{R}$ such that

(1)
$$\{\omega \in \Omega : X(\omega) \leq x\} \in \mathscr{F} \quad \text{for all } x \in \mathbb{R}.$$

We require that random variables satisfy (1) for very much the same reason as we required (2.2) for discrete random variables. That is, we are interested in the values taken by a random variable X and the likelihoods of these values. It turns out that the right way to do this is to fix $x \in \mathbb{R}$ and ask for the probability that X takes a value in $(-\infty, x]$; this probability exists only if the inverse image $X^{-1}((-\infty, x]) = \{\omega \in \Omega : X(\omega) \leq x\}$ of $(-\infty, x]$ lies in the event space $\mathscr{F}$, and so we postulate that this is the case for all $x \in \mathbb{R}$. Note that every discrete random variable X is a random variable. To see this, observe that if X is a discrete random variable then

$$\{\omega \in \Omega : X(\omega) \leq x\} = \bigcup_{\substack{y \in \mathrm{Im}X : \\ y \leq x}} \{\omega \in \Omega : X(\omega) = y\},$$

which is the countable union of events in $\mathscr{F}$ and therefore belongs to $\mathscr{F}$.

Whereas discrete random variables were studied through their *mass* functions, random variables in the broader sense are studied through their *distribution* functions, defined as follows. If X is a random variable on $(\Omega, \mathscr{F}, \mathsf{P})$, the *distribution function* of X is a mapping $F_X : \mathbb{R} \to [0, 1]$ defined by

(2)
$$F_X(x) = \mathsf{P}(\{\omega \in \Omega : X(\omega) \leq x\}).$$

We often write (2) as '$F_X(x) = \mathsf{P}(X \leq x)$'.

Example 3
Suppose that X is a discrete random variable taking non-negative integer values, with mass function

$$\mathsf{P}(X = k) = p(k) \qquad \text{for } k = 0, 1, 2, \ldots.$$

For any $x \in \mathbb{R}$ it is the case that $X \leq x$ if and only if X takes one of the values $0, 1, 2, \ldots, \lfloor x \rfloor$, where $\lfloor x \rfloor$ denotes the greatest integer not greater than x. Hence

$$F_X(x) = \begin{cases} 0 & \text{if } x < 0, \\ p(0) + p(1) + \cdots + p(\lfloor x \rfloor) & \text{if } x \geq 0, \end{cases}$$

and a sketch of this function is displayed in Fig. 5.1. $\qquad\square$

The distribution function F_X of a random variable X has various general and elementary properties, the first of which is

(4)
$$F_X(x) \leq F_X(y) \qquad \text{if } x \leq y,$$

which is to say that F_X is monotonic non-decreasing. This holds because

$$\{\omega \in \Omega : X(\omega) \leq x\} \subseteq \{\omega \in \Omega : X(\omega) \leq y\}$$

whenever $x \leq y$, since if X takes a value smaller than x then this value is certainly smaller than y. Other elementary properties of $F_X(x)$ concern its behaviour when x is near $-\infty$ or $+\infty$. It is intuitively clear that

(5)
$$F_X(x) \to 0 \qquad \text{as } x \to -\infty,$$

and

(6)
$$F_X(x) \to 1 \qquad \text{as } x \to \infty,$$

since in the first case, as $x \to -\infty$ the event that X is smaller than x becomes less and less likely, whilst in the second case, as $x \to \infty$ this event becomes overwhelmingly likely. At an intuitive level (5) and (6) are obvious, since they resemble the trivial remarks

$$P(X \leq -\infty) = 0, \qquad P(X \leq \infty) = 1,$$

but a formal verification of (5) and (6) relies on Theorem 1C. In the

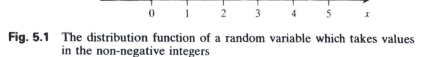

Fig. 5.1 The distribution function of a random variable which takes values in the non-negative integers

same way, Theorem 1C is needed to prove the last general property of distribution functions:

(7) $$F_X \text{ is continuous from the right,}$$

which is to say that†

(8) $$F_X(x + \varepsilon) \to F_X(x) \qquad \text{as } \varepsilon \downarrow 0.$$

A glance at Fig. 5.1 confirms that distribution functions need not be continuous from the left. Properties (4)–(7) characterize distribution functions completely, in the sense that if F is a function which satisfies (4)–(7) then there exists a probability space and a random variable X on this space such that X has distribution function F. The proof of this result is too deep to be given at this level, but this fact should be noted since in many circumstances it allows us to avoid the rather tedious business of writing down probability spaces and random variables explicitly.

Before we move on to give examples of distribution functions, here is a final property. The probability $F_X(x) = P(X \le x)$ is the probability that X takes a value in the infinite interval $(-\infty, x]$. To find the probability that X takes a value in the bounded interval $(a, b]$ we proceed in the following way. Let $a < b$; then

$$P(a < X \le b) = P(\{X \le b\} \backslash \{X \le a\})$$
$$= P(X \le b) - P(X \le a)$$

since the event $\{X \le a\}$ is a subset of the event $\{X \le b\}$. Hence

(9) $$P(a < X \le b) = F_X(b) - F_X(a).$$

Exercises

1. Let X be a random variable taking integral values such that $P(X = k) = p(k)$ for $k = \ldots, -1, 0, 1, \ldots$. Show that the distribution function of X satisfies

$$F_X(b) - F_X(a) = p(a + 1) + p(a + 2) + \cdots + p(b)$$

for all integers a, b with $a < b$.

2. If X is a random variable and c is a real number such that $P(X = c) > 0$, show that the distribution function $F_X(x)$ of X is discontinuous at the point $x = c$. Is the converse true?

3. If X has distribution function $F_X(x)$, what is the distribution function of $Y = \max\{0, X\}$?

4. The real number m is called a *median* of the random variable X if

$$P(X < m) \le \tfrac{1}{2} \le P(X \le m).$$

Show that every random variable has at least one median.

---

† The limit in (8) is taken as ε tends down to 0 through positive values only.

5.2   Examples of distribution functions

Example 3 contains our first example of a distribution function. Note the general features of this function: non-decreasing, continuous from the right, tending to 0 as $x \to -\infty$ and to 1 as $x \to \infty$. Other distribution functions contrast starkly to this function by being continuous, and our next example is such a function.

Example 10

Uniform distribution. Let $a, b \in \mathbb{R}$ and $a < b$. The function

(11)
$$F(x) = \begin{cases} 0 & \text{if } x < a \\ \dfrac{x-a}{b-a} & \text{if } a \leq x \leq b, \\ 1 & \text{if } x > b, \end{cases}$$

sketched in Fig. 5.2, has the properties (4)–(7) and is thus a distribution function. A random variable with this distribution function is said to have the *uniform distribution* on the interval (a, b); some people call this the uniform distribution on $[a, b]$.   □

Example 12

Exponential distribution. Let $\lambda > 0$ and let F be given by

(13)
$$F(x) = \begin{cases} 0 & \text{if } x \leq 0, \\ 1 - e^{-\lambda x} & \text{if } x > 0, \end{cases}$$

see Fig. 5.3 for a sketch of this function. Clearly F is a distribution function. A random variable with this distribution is said to have the *exponential distribution* with parameter λ.   □

The two distribution functions above are very important in probability theory. There are many other distribution functions of course; for example, any non-negative function F which is continuous and non-decreasing and satisfies

$$\lim_{x \to -\infty} F(x) = 0, \qquad \lim_{x \to \infty} F(x) = 1$$

is a distribution function.

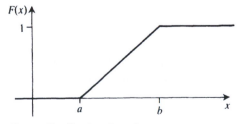

Fig. 5.2   The uniform distribution function

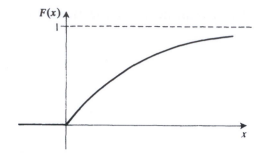

Fig. 5.3   The exponential distribution function

Exercises   5. Show that if F_1 and F_2 are distribution functions, then so is $F(x) =$ $\alpha F_1(x) + (1 - \alpha)F_2(x)$ for any α satisfying $0 \le \alpha \le 1$.
6. Let

$$F(x) = c \int_{-\infty}^{x} e^{-|u|} \, du \qquad \text{for } x \in \mathbb{R}.$$

For what value of c is F a distribution function?

5.3   Continuous random variables

Random variables come in many shapes, but there are two classes of random variable which are particularly important:

I: discrete random variables,

II: continuous random variables.

Discrete random variables take only countably many values and their distribution functions generally look like step functions (remember Fig. 5.1). At the other extreme, there are random variables whose distribution functions are very smooth (remember Figs. 5.2, 5.3) and we call such random variables 'continuous'. More formally, we call a random variable X *continuous*† if its distribution function F_X may be written in the form

(14) $$F_X(x) = P(X \le x) = \int_{-\infty}^{x} f_X(u) \, du \qquad \text{for all } x \in \mathbb{R},$$

for some non-negative function f_X. In this case we say that X has *(probability) density function* f_X.

---

† More advanced textbooks call such random variables 'absolutely continuous'.

Example 15
If X has the exponential distribution with parameter λ then

$$F_X(x) = \begin{cases} 0 & \text{if } x \le 0, \\ 1 - e^{-\lambda x} & \text{if } x > 0, \end{cases}$$

and

$$f_X(x) = \begin{cases} 0 & \text{if } x \le 0, \\ \lambda e^{-\lambda x} & \text{if } x > 0. \end{cases} \qquad \square$$

Provided that X is a continuous random variable and F_X is well-behaved in (14), we can take

(16)
$$f_X(x) = \begin{cases} \dfrac{d}{dx} F_X(x) & \text{if this derivative exists at } x, \\ 0 & \text{otherwise}, \end{cases}$$

as the density function of X. We shall normally do this, although we should point out that there are some difficulties over mathematical rigour here. However, for almost all practical purposes (16) is adequate, and the reader of a text at this level should seldom get into trouble if he uses (16) when finding density functions of continuous random variables.

Density functions serve continuous random variables in very much the same way as mass functions serve discrete random variables, and it is not surprising that the general properties of density functions and mass functions are very similar. For example, it is clear that the density function f_X of X satisfies

(17)
$$f_X(x) \ge 0 \qquad \text{for all } x \in \mathbb{R}, \qquad (p_Y(x) \ge 0)$$

(18)
$$\int_{-\infty}^{\infty} f_X(x)\, dx = 1, \qquad \left(\sum_x p_Y(x) = 1 \right)$$

where the parentheses contain the corresponding properties, (2.4) and (2.5), of a mass function p_Y. However, this analogy can be dangerous, since $f_X(x)$ is *not* a probability and may well even exceed 1 in value. On the other hand, $f_X(x)$ is indeed a 'measure' of probability in the following sense. If δx is small and positive then, roughly speaking, the probability that X is 'near' to x satisfies

(19)
$$P(x < X \le x + \delta x) = F(x + \delta x) - F(x) \qquad \text{by (9)}$$
$$= \int_x^{x + \delta x} f_X(u)\, du \qquad \text{by (14)}$$
$$\approx f_X(x)\, \delta x \qquad \text{for small } \delta x.$$

So the true analogy is not between a density function $f_X(x)$ and a mass function $p_Y(x)$, but between $f_X(x)\,\delta x$ and $p_Y(x)$. This is borne out by comparing (18) with (2.5): values of the mass function are replaced by $f_X(x)\,\delta x$ and the summation (since, for discrete random variables, only countably many values are positive) is replaced by the integral. A startling difference between discrete and continuous random variables is given in the first part of the next theorem.

Theorem 5A *If X is continuous with density function f_X then*

(20)
$$P(X = x) = 0 \quad \text{for all } x \in \mathbb{R},$$

(21)
$$P(a \le X \le b) = \int_a^b f_X(u)\,du \quad \text{for all } a, b \in \mathbb{R} \text{ with } a \le b.$$

Proof We argue as follows:

$$P(X = x) = \lim_{\varepsilon \downarrow 0} P(x - \varepsilon < X \le x)$$

$$= \lim_{\varepsilon \downarrow 0} (F_X(x) - F_X(x - \varepsilon)) \qquad \text{by (9)}$$

$$= \lim_{\varepsilon \downarrow 0} \int_{x-\varepsilon}^x f_X(u)\,du \qquad \text{by (14)}$$

$$= 0.$$

The first equality here cannot be justified without an appeal to the continuity of P (Theorem 1C), but we do not spell this out. For the second part of the theorem, if $a \le b$ then

$$P(a \le X \le b) = P(a < X \le b) \qquad \text{by (20)}$$

$$= F_X(b) - F_X(a) \qquad \text{by (9)}$$

$$= \int_a^b f_X(u)\,du. \qquad \square$$

To recap, all random variables have a distribution function, and in addition discrete random variables have a mass function and continuous random variables have a density function. There are many random variables which are neither discrete nor continuous, and we shall come across some of these later.

Exercises 7. A random variable X has density function

$$f(x) = \begin{cases} 2x & \text{if } 0 < x < 1, \\ 0 & \text{otherwise.} \end{cases}$$

Find the distribution function of X.

8. If X has density function

$$f(x) = \tfrac{1}{2} e^{-|x|} \qquad \text{for } x \in \mathbb{R},$$

find the distribution function of X. This is called the *bilateral* (or *double*) *exponential* distribution.

9. If X has distribution function

$$F(x) = \begin{cases} \dfrac{1}{2(1+x^2)} & \text{if } -\infty < x \le 0, \\[2mm] \dfrac{1+2x^2}{2(1+x^2)} & \text{if } 0 < x < \infty, \end{cases}$$

show that X is continuous and find its density function.

10. Find the distribution function of the so-called 'extreme value' density function

$$f(x) = \exp(-x - e^{-x}) \qquad \text{for } x \in \mathbb{R}.$$

5.4 Some common density functions

It is fairly clear that any function f which satisfies

(22) $$f(x) \ge 0 \qquad \text{for all } x,$$

and

(23) $$\int_{-\infty}^{\infty} f(x) \, dx = 1,$$

is the density function of some random variable. To confirm this, simply define

$$F(x) = \int_{-\infty}^{x} f(u) \, du$$

and check that F is a distribution function by verifying (4)–(7). There are a few such functions f which are important in practice, and we list these below.

The *uniform distribution* on (a, b) has density function

(24) $$f(x) = \begin{cases} (b-a)^{-1} & \text{if } a < x < b, \\ 0 & \text{otherwise.} \end{cases}$$

The *exponential distribution* with parameter $\lambda > 0$ has density function

(25) $$f(x) = \begin{cases} \lambda e^{-\lambda x} & \text{if } x > 0, \\ 0 & \text{if } x \le 0. \end{cases}$$

The *normal distribution* with parameters μ and σ^2 (sometimes written

$N(\mu, \sigma^2))$ has density function

(26)
$$f(x) = \frac{1}{\sqrt{(2\pi\sigma^2)}} \exp\left(-\frac{1}{2\sigma^2}(x-\mu)^2\right) \quad \text{for } -\infty < x < \infty.$$

The *Cauchy distribution* has density function

(27)
$$f(x) = \frac{1}{\pi(1+x^2)} \quad \text{for } -\infty < x < \infty.$$

The *gamma distribution* with parameters w (>0) and λ (>0) has density function

(28)
$$f(x) = \begin{cases} \dfrac{1}{\Gamma(w)} \lambda^w x^{w-1} e^{-\lambda x} & \text{if } x > 0, \\ 0 & \text{if } x \le 0, \end{cases}$$

where $\Gamma(w)$ is the gamma function, defined by

(29)
$$\Gamma(w) = \int_0^\infty x^{w-1} e^{-x} \, dx.$$

Note that, for positive integers w, $\Gamma(w) = (w-1)!$.
For $n = 1, 2, \ldots$, the *chi-squared* (χ^2) *distribution with n degrees of freedom* (sometimes written χ_n^2) has density function

(30)
$$f(x) = \begin{cases} \dfrac{1}{2\Gamma(\frac{1}{2}n)} (\tfrac{1}{2}x)^{\frac{1}{2}n-1} e^{-\frac{1}{2}x} & \text{if } x > 0, \\ 0 & \text{if } x \le 0. \end{cases}$$

Comparison of (30) with (28) shows that the χ_n^2 distribution is the same as the gamma distribution with parameters $\frac{1}{2}n$ and $\frac{1}{2}$, but we list the distribution separately here because of its common occurrence in statistical analysis.

This list is a dull compendium of some of the commoner density functions, and we do not expect it to inspire the reader in this form. It is difficult to motivate these density functions adequately at this stage, but we shall need to refer back to this section later when we meet these functions in action.

It is not always a trivial task to show that these functions are actually density functions. The condition (22) of non-negativity is no problem, but some care is required in checking that they integrate to 1. For example, to check this for the function given in (26) we require the standard definite integral

$$\int_{-\infty}^\infty e^{-x^2} \, dx = \sqrt{\pi}.$$

Of course, the constant terms in (24)–(30) have been chosen solely so that the resulting function integrates to 1. For example, it is clear that the function

$$g(x) = 1/(1 + x^2) \qquad \text{for } -\infty < x < \infty.$$

is not a density function since

$$\int_{-\infty}^{\infty} g(x) \, dx = \pi,$$

but it follows that the 'normalized' function

$$f(x) = (1/\pi)g(x)$$

is a density function.

Exercises   11. For what values of its parameters is the gamma distribution also an exponential distribution?

12. Show that the gamma function $\Gamma(w)$ satisfies $\Gamma(w) = (w - 1)\Gamma(w - 1)$ for all $w > 1$, and deduce that $\Gamma(n) = (n - 1)!$ for $n = 1, 2, 3, \ldots$.

13. Show that the density function

$$f(x) = \begin{cases} \dfrac{1}{\pi\sqrt{[x(1-x)]}} & \text{if } 0 < x < 1, \\ 0 & \text{otherwise}, \end{cases}$$

has distribution function with the form

$$F(x) = c \arcsin \sqrt{x} \qquad \text{if } 0 \le x \le 1$$

and find the constant c.

5.5   Functions of random variables

Let X be a random variable on $(\Omega, \mathcal{F}, \mathsf{P})$ and suppose that $g : \mathbb{R} \to \mathbb{R}$. Then $Y = g(X)$ is a mapping from Ω into $\mathbb{R}$, defined by $Y(\omega) = g[X(\omega)]$ for $\omega \in \Omega$. Actually Y is not generally a random variable since it need not satisfy condition (1). It turns out, however, that (1) is valid for Y whenever g is sufficiently well behaved (such as g is a continuous function, or a monotone function, or . . .) and so we neglect this difficulty, assuming henceforth that *all quantities of the form $Y = g(X)$ are random variables*. The main question is now the following: if we know the distribution of X, then how do we find the distribution of $Y = g(X)$? If X is discrete with mass function p_X then (2.18) provides the answer, and we consider next the case when X is continuous with density function f_X. We begin with an example.

Example 31

If X is continuous with density function f_X, and $g(x) = ax + b$ where $a > 0$, then $Y = g(X) = aX + b$ has distribution function given by

$$P(Y \le y) = P(aX + b \le y)$$
$$= P(X \le a^{-1}(y - b))$$
$$= F_X(a^{-1}(y - b)),$$

and differentiation with respect to y yields

$$f_Y(y) = a^{-1}f_X(a^{-1}(y - b)) \qquad \text{for all } y \in \mathbb{R}. \qquad \square$$

The next theorem generalizes the result of this example.

Theorem 5B

If X is a continuous random variable with density function f_X, and g is a strictly increasing and differentiable function from $\mathbb{R}$ into $\mathbb{R}$, then $Y = g(X)$ has density function

(32)
$$f_Y(y) = f_X(g^{-1}(y)) \frac{d}{dy} [g^{-1}(y)] \qquad \text{for } y \in \mathbb{R},$$

where g^{-1} is the inverse function of g.

Proof

First we find the distribution function of Y:

$$P(Y \le y) = P(g(X) \le y)$$
$$= P(X \le g^{-1}(y)) \text{ since } g \text{ is increasing.}$$

We differentiate this with respect to y, noting that the right-hand side is a function of a function, to obtain (32). $\qquad \square$

If, in Theorem 5B, g were strictly *decreasing* then the same argument gives that $Y = g(X)$ has density function

(33)
$$f_Y(y) = -f_X(g^{-1}(y)) \frac{d}{dy} [g^{-1}(y)] \qquad \text{for } y \in \mathbb{R}.$$

Formulae (32) and (33) rely heavily on the monotonicity of g; other cases are best treated on their own merits, and actually there is a lot to be said for using the method of the next example always, rather than taking recourse in the general results (32)–(33).

Example 34

If X has density function f_X, and $g(x) = x^2$, then $Y = g(X) = X^2$ has distribution function

$$P(Y \le y) = P(X^2 \le y)$$
$$= \begin{cases} 0 & \text{if } y < 0, \\ P(-\sqrt{y} \le X \le \sqrt{y}) & \text{if } y \ge 0. \end{cases}$$

Hence $f_Y(y) = 0$ if $y \leq 0$, while for $y > 0$

$$f_Y(y) = \frac{d}{dy} P(Y \leq y) \qquad \text{if this derivative exists}$$

$$= \frac{d}{dy} [F_X(\sqrt{y}) - F_X(-\sqrt{y})]$$

$$= \frac{1}{2\sqrt{y}} [f_X(\sqrt{y}) + f_X(-\sqrt{y})]. \qquad \square$$

Exercises   14. Let X be a random variable with the exponential distribution, parameter λ. Find the density function of
  (i) $A = 2X + 5$,   (ii) $B = e^x$,
  (iii) $C = (1 + X)^{-1}$,   (iv) $D = (1 + X)^{-2}$.
15. Show that if X has the normal distribution with parameters 0 and 1, then $Y = X^2$ has the χ^2 distribution with one degree of freedom.

5.6   Expectations of continuous random variables

If a one-dimensional rod of metal has density $\rho(x)$ at point x, then its centre of gravity is at $\int x\rho(x)\, dx / \int \rho(x)\, dx$. This leads naturally to the idea of the expectation of a continuous random variable.

If X is a continuous random variable with density function f_X, then the *expectation* of X is denoted by $E(X)$ and defined by

(35) $$E(X) = \int_{-\infty}^{\infty} x f_X(x)\, dx,$$

whenever this integral converges absolutely (in that $\int_{-\infty}^{\infty} |x f_X(x)|\, dx < \infty$).

As in the case of discrete variables, the expectation of X is often called the *expected value* or *mean* of X.

If X is a continuous variable and $g : \mathbb{R} \to \mathbb{R}$ then $Y = g(X)$ is a random variable also (so long as g is sufficiently well behaved). It may be difficult to calculate $E(Y)$ from first principles, not least since Y may be neither discrete nor continuous and so neither of formulae (2.19) and (35) may apply. Of great value here is the following result, which enables us to calculate $E(Y)$ directly from knowledge of f_X and g.

Theorem 5C   *If X is a continuous random variable with density function f_X, and $g : \mathbb{R} \to \mathbb{R}$, then*

(36) $$E(g(X)) = \int_{-\infty}^{\infty} g(x) f_X(x)\, dx,$$

whenever this integral converges absolutely.

Sketch proof
Theorem 5C is not too difficult to prove, but the proof is a little long (see Grimmett and Stirzaker (1982, pp. 51–52) for a discussion). We think that it is more important to understand *why* it is true rather than to see a proof. If Y is a discrete random variable and $g : \mathbb{R} \to \mathbb{R}$ then

(37)
$$E(g(Y)) = \sum_x g(x)P(Y = x);$$

this is the conclusion of Theorem 2B. Remember the analogy between mass functions and density functions: in (37), replace $P(Y = x)$ by $f_X(x)\,dx$ and the summation by the integral, to obtain (36). See also Problem 8 at the end of this chapter. □

As in the case of discrete random variables, the mean $E(X)$ of a continuous random variable X is an indication of the 'centre' of the distribution of X. As a measure of the degree of dispersion of X about this mean we normally take the *variance* of X, defined to be

(38)
$$\text{var}(X) = E([X - E(X)]^2).$$

We have from Theorem 5C that

$$\text{var}(X) = \int_{-\infty}^{\infty} (x - \mu)^2 f_X(x)\,dx$$

where $\mu = E(X)$, and so

$$\text{var}(X) = \int_{-\infty}^{\infty} (x^2 - 2\mu x + \mu^2) f_X(x)\,dx$$

$$= \int_{-\infty}^{\infty} x^2 f_X(x)\,dx - 2\mu \int_{-\infty}^{\infty} x f_X(x)\,dx + \mu^2 \int_{-\infty}^{\infty} f_X(x)\,dx$$

$$= \int_{-\infty}^{\infty} x^2 f_X(x)\,dx - \mu^2$$

by (35) and (23). Thus we obtain the usual formula

(39)
$$\text{var}(X) = E(X^2) - E(X)^2.$$

Example 40
If X has the uniform distribution on (a, b) then

$$E(X) = \int_{-\infty}^{\infty} x f_X(x)\,dx \qquad \text{by (35)}$$

$$= \int_{a}^{b} x \frac{1}{b - a}\,dx$$

$$= \tfrac{1}{2}(a + b).$$

As an example of a function of a random variable, let $Y = \sin X$.

Then

$$E(Y) = \int_{-\infty}^{\infty} \sin x \, f_X(x) \, dx \qquad \text{by (36)}$$

$$= \int_a^b \sin x \, \frac{1}{b-a} \, dx$$

$$= \frac{\cos a - \cos b}{b-a}. \qquad \square$$

Example 41   If X has the exponential distribution with parameter λ then

$$E(X) = \int_0^\infty x\lambda e^{-\lambda x} \, dx = \frac{1}{\lambda},$$

and

$$E(X^2) = \int_0^\infty x^2 \lambda e^{-\lambda x} \, dx = \frac{2}{\lambda^2}$$

by (36), giving by (39) that the variance of X is

$$\text{var}(X) = E([X - E(X)]^2)$$

$$= E(X^2) - E(X)^2 = \frac{1}{\lambda^2}. \qquad \square$$

Example 42   If X has the normal distribution with parameters $\mu = 0$ and $\sigma^2 = 1$ then

$$E(X) = \int_{-\infty}^{\infty} x \, \frac{1}{\sqrt{(2\pi)}} \, e^{-\frac{1}{2}x^2} \, dx = 0,$$

by symmetry properties of the integrand. Hence

$$\text{var}(X) = \int_{-\infty}^{\infty} x^2 \, \frac{1}{\sqrt{(2\pi)}} \, e^{-\frac{1}{2}x^2} \, dx = 1.$$

Similar integrations show that the normal distribution with parameters μ and σ^2 has mean μ and variance σ^2, as we may have expected. $\square$

Example 43   If X has the Cauchy distribution then $E(X)$ is given by

$$E(X) = \int_{-\infty}^{\infty} x \, \frac{1}{\pi(1+x^2)} \, dx,$$

so long as this integral exists. *It does not exist*, since

$$\int_{-M}^{N} x \frac{1}{\pi(1+x^2)} \, dx = \left[\frac{1}{2\pi} \log(1+x^2)\right]_{-M}^{N}$$

$$= \frac{1}{2\pi} \log \frac{1+N^2}{1+M^2}$$

$$= l(M, N),$$

say, and the limit of $l(M, N)$ as $M, N \to \infty$ depends on the way in which M and N approach ∞: if $M \to \infty$ and $N \to \infty$ in that order, then $l(M, N) \to -\infty$, while if the limit is taken in the other order then $l(M, N) \to \infty$. Hence *the Cauchy distribution does not have a mean value*. On the other hand there are many functions of X with finite expectations. For example, if $Y = \arctan X$ then

$$E(Y) = \int_{-\infty}^{\infty} (\arctan x) \frac{1}{\pi(1+x^2)} \, dx$$

$$= \int_{-\frac{1}{2}\pi}^{\frac{1}{2}\pi} \frac{v}{\pi} \, dv \qquad \text{(where } v = \arctan x)$$

$$= 0. \qquad \qquad \square$$

Exercises 16. Show that the mean value of a random variable with density function

$$f(x) = \begin{cases} \dfrac{1}{\pi\sqrt{[x(1-x)]}} & \text{if } 0 < x < 1, \\ 0 & \text{otherwise,} \end{cases}$$

is $\frac{1}{2}$.

17. If X has the normal distribution with mean 0 and variance 1, find the mean value of $Y = e^{2X}$.

5.7 Problems

1. The *bilateral* (or *double*) *exponential* distribution has density function

$$f(x) = \tfrac{1}{2} c e^{-c|x|} \qquad \text{for } x \in \mathbb{R},$$

where c (>0) is a parameter of the distribution. Show that the mean and variance of this distribution are 0 and $2c^{-2}$ respectively.

2. Let X be a random variable with the Poisson distribution, parameter λ. Show that, for $w = 1, 2, 3, \ldots$,

$$P(X \geq w) = P(Y \leq \lambda)$$

where Y is a random variable having the gamma distribution with parameters w and 1.

3. The random variable X has density function proportional to $g(x)$ where g

is a function satisfying

$$g(x) = \begin{cases} |x|^{-n} & \text{if } |x| \geq 1, \\ 0 & \text{otherwise,} \end{cases}$$

and n (≥ 2) is an integer. Find and sketch the density function of X, and determine the values of n for which both the mean and variance of X exist.

4. If X has the normal distribution with mean 0 and variance 1, find the density function of $Y = |X|$, and find the mean and variance of Y.

5. Let X be a random variable whose distribution function F is a continuous function. Show that the random variable Y, defined by $Y = F(X)$, is uniformly distributed on the interval $(0, 1)$.

*6. Let F be a distribution function, and let X be a random variable which is uniformly distributed on the interval $(0, 1)$. Let F^{-1} be the inverse function of F, defined by

$$F^{-1}(y) = \inf\{x : F(x) \geq y\},$$

and show that the random variable $Y = F^{-1}(X)$ has distribution function F. (This observation may be used in practice to generate pseudo-random numbers from any given distribution.)

7. If X is a continuous random variable taking non-negative values only, show that

$$E(X) = \int_0^\infty [1 - F_X(x)]\, dx,$$

whenever this integral exists.

*8. Use the result of Problem 7 to show that

$$E(g(X)) = \int_{-\infty}^\infty g(x) f_X(x)\, dx$$

whenever X and $g(X)$ are continuous random variables and g is a non-negative function on $\mathbb{R}$.

9. The random variable X' is said to be obtained from the random variable X by 'truncation at the point a' if X' is defined by

$$X'(\omega) = \begin{cases} X(\omega) & \text{if } X(\omega) \leq a, \\ a & \text{if } X(\omega) > a. \end{cases}$$

Express the distribution function of X' in terms of the distribution function of X.

10. Let X be uniformly distributed on the interval $(1, 5)$. Find the distribution and density functions of $Y = X/(5 - X)$.   (Bristol 1984)

11. William Tell is a very bad shot. In practice, he places a small green apple on top of a straight wall which stretches to infinity in both directions. He then takes up position at a distance of one perch from the apple, so that his line of sight to the target is perpendicular to the wall. He now selects an angle uniformly at random from his entire field of view and shoots his arrow in this direction. Assuming that his arrow hits the wall somewhere, what is the distribution function of the horizontal distance (measured in perches) between the apple and the point which the arrow strikes. There is no wind.

B. Further Probability

6

Multivariate distributions and independence

6.1 Random vectors and independence

Given two random variables, X and Y, acting on a probability space $(\Omega, \mathscr{F}, \mathsf{P})$, it is often necessary to think of them acting together as a random vector (X, Y) taking values in $\mathbb{R}^2$. If X and Y are discrete, we may study this random vector by using the joint mass function of X and Y, but this method is not always available. In the general case of arbitrary random variables X, Y we study instead the *joint distribution function* of X and Y, defined as the mapping $F_{X,Y} : \mathbb{R}^2 \to [0, 1]$ given by

$$(1) \qquad F_{X,Y}(x, y) = \mathsf{P}(\{\omega \in \Omega : X(\omega) \leq x,\ Y(\omega) \leq y\})$$
$$= \mathsf{P}(X \leq x,\ Y \leq y).$$

Joint distribution functions have certain elementary properties which are exactly analogous to those of ordinary distribution functions. For example, it is easy to see that

$$(2) \qquad \lim_{\substack{x \to -\infty \\ y \to -\infty}} F_{X,Y}(x, y) = 0,$$

and

$$(3) \qquad \lim_{\substack{x \to \infty \\ y \to \infty}} F_{X,Y}(x, y) = 1,$$

just as in (5.5) and (5.6). Similarly, $F_{X,Y}$ is non-increasing in each variable in that

$$(4) \qquad F_{X,Y}(x_1, y_1) \leq F_{X,Y}(x_2, y_2) \quad \text{if} \quad x_1 \leq x_2 \text{ and } y_1 \leq y_2.$$

The joint distribution function $F_{X,Y}$ contains a great deal more information than the two ordinary distribution functions F_X and F_Y, since it tells us how X and Y behave *together*. In particular, the distribution functions of X and Y may be found from their joint distribution function in a routine way. It is intuitively attractive to write

$$F_X(x) = \mathsf{P}(X \leq x)$$
$$= \mathsf{P}(X \leq x,\ Y \leq \infty) = F_{X,Y}(x, \infty)$$

and similarly

$$F_Y(y) = F_{X,Y}(\infty, y),$$

but the mathematically correct way of expressing this is

(5)
$$F_X(x) = \lim_{y \to \infty} F_{X,Y}(x, y)$$

and

(6)
$$F_Y(y) = \lim_{x \to \infty} F_{X,Y}(x, y).$$

These distribution functions are called the *marginal* distribution functions of the joint distribution function $F_{X,Y}$.

The idea of the 'independence' of random variables X and Y follows naturally from this discussion: we call X and Y *independent* if the pair, $\{\omega \in \Omega : X(\omega) \leq x\}$ and $\{\omega \in \Omega : Y(\omega) \leq y\}$, are independent events for all $x, y \in \mathbb{R}$. That is to say, X and Y are independent if and only if

$$P(X \leq x, Y \leq y) = P(X \leq x)P(Y \leq y) \qquad \text{for } x, y \in \mathbb{R},$$

which is to say that their joint distribution function factorizes as the product of the two marginal distribution functions:

(7)
$$F_{X,Y}(x, y) = F_X(x)F_Y(y) \qquad \text{for } x, y \in \mathbb{R}.$$

It is a straightforward exercise to show that this is a genuine extension of the notion of independent *discrete* random variables. Random variables which are not independent are called *dependent*.

We study families of random variables in very much the same way. Briefly, if $X_1, X_2, \ldots, X_n$ are random variables on $(\Omega, \mathcal{F}, P)$, their *joint distribution function* is a function $F_X : \mathbb{R}^n \to [0, 1]$ given by

(8)
$$F_X(x) = P(X_1 \leq x_1, X_2 \leq x_2, \ldots, X_n \leq x_n)$$

for all $x = (x_1, x_2, \ldots, x_n) \in \mathbb{R}^n$. The variables $X_1, X_2, \ldots, X_n$ are called *independent* if

$$P(X_1 \leq x_1, \ldots, X_n \leq x_n) = P(X_1 \leq x_1) \cdots P(X_n \leq x_n) \qquad \text{for all } x \in \mathbb{R}^n,$$

or equivalently if

(9)
$$F_X(x) = F_{X_1}(x_1) \cdots F_{X_n}(x_n) \qquad \text{for all } x \in \mathbb{R}^n.$$

Example 10 Suppose that X and Y are random variables on some probability space, each taking values in the integers $\{\ldots, -1, 0, 1, \ldots\}$ with joint mass function

$$P(X = i, Y = j) = p(i, j) \qquad \text{for } i, j = 0, \pm 1, \pm 2, \ldots.$$

Their joint distribution function is given by

$$F_{X,Y}(x, y) = \sum_{\substack{i \le x \\ j \le y}} p(i, j) \qquad \text{for } (x, y) \in \mathbb{R}^2. \qquad \square$$

Example 11

Suppose that X and Y are random variables with joint distribution function

$$F_{X,Y}(x, y) = \begin{cases} 1 - e^{-x} - e^{-y} + e^{-x-y} & \text{if } x, y \ge 0, \\ 0 & \text{otherwise.} \end{cases}$$

The (marginal) distribution function of X is

$$F_X(x) = \lim_{y \to \infty} F_{X,Y}(x, y)$$

$$= \begin{cases} 1 - e^{-x} & \text{if } x \ge 0, \\ 0 & \text{otherwise,} \end{cases}$$

so that X has the exponential distribution with parameter 1. A similar calculation shows that Y has this distribution also. Hence

$$F_{X,Y}(x, y) = F_X(x)F_Y(y) \qquad \text{for } x, y \in \mathbb{R},$$

and so X and Y are independent. $\qquad \square$

Exercises

1. Show that two random variables X and Y are independent if and only if

$$P(X > x, Y > y) = P(X > x)P(Y > y) \qquad \text{for all } x, y \in \mathbb{R}.$$

2. Let the pair (X, Y) of random variables have joint distribution function $F(x, y)$. Prove that

$$P(a < X \le b, c < Y \le d) = F(b, d) + F(a, c) - F(a, d) - F(b, c)$$

for any $a, b, c, d \in \mathbb{R}$ such that $a < b$ and $c < d$.

3. Prove that two random variables X and Y are independent if and only if

$$P(a < X \le b, c < Y \le d) = P(a < X \le b)P(c < Y \le d)$$

for all $a, b, c, d \in \mathbb{R}$ satisfying $a < b$ and $c < d$.

6.2 Joint density functions

We have called a random variable X 'continuous' if its distribution function may be expressed in the form

$$F_X(x) = P(X \le x) = \int_{-\infty}^{x} f(u) \, du \qquad \text{for } x \in \mathbb{R}.$$

In the same way, the pair X, Y of random variables on $(\Omega, \mathcal{F}, P)$ is called (*jointly*) *continuous* if its joint distribution function is express-ible in the form†

(12)
$$F_{X,Y}(x, y) = P(X \leq x, Y \leq y) = \int_{u=-\infty}^{x} \int_{v=-\infty}^{y} f(u, v) \, du \, dv$$

for all $x, y \in \mathbb{R}$ and some function $f : \mathbb{R}^2 \to [0, \infty)$. If this holds, then we say that X and Y have *joint (probability) density function f*, and we usually denote this function by $f_{X,Y}$.

As in Section 5.3, if X and Y are jointly continuous then we may take their joint density function to be that function given by

(13)
$$f_{X,Y}(x, y) = \begin{cases} \dfrac{\partial^2}{\partial x \, \partial y} F_{X,Y}(x, y) & \text{if this derivative exists at } (x, y), \\ 0 & \text{otherwise,} \end{cases}$$

and we shall normally do this in future. There are the usual problems here over mathematical rigour but, as noted after (5.16), you should not get into trouble at this level if you take (13) to be the definition of the joint density function of X and Y.

The elementary properties of the joint density function $f_{X,Y}$ are consequences of properties (2)–(4) of joint distribution functions:

(14)
$$f_{X,Y}(x, y) \geq 0 \qquad \text{for all } x, y \in \mathbb{R},$$

and

(15)
$$\int_{-\infty}^{\infty} \int_{-\infty}^{\infty} f_{X,Y}(x, y) \, dx \, dy = 1.$$

Once again we note an analogy between joint density functions and joint mass functions. This may be expressed rather crudely by saying that for any $(x, y) \in \mathbb{R}^2$ and small positive δx and δy, the probability that the random vector (X, Y) lies in the small rectangle with bottom left-hand corner at (x, y) and side-lengths δx and δy is

(16)
$$P(x < X \leq x + \delta x, \, y < Y \leq y + \delta y) \approx f_{X,Y}(x, y) \, \delta x \, \delta y;$$

see Fig. 6.1 for a diagram of this region. This holds for very much the same reasons as the one-dimensional case (5.19). It is not difficult to see how this leads to the next theorem.

† We ought to say exactly what we mean by the integral on the right-hand side of (12). At this level, it is perhaps enough to say that this double integral may be interpreted in any standard way, and that there is a result (called Fubini's Theorem) which says that, under certain very wide conditions, it does not matter whether we integrate over u first or over v first when we calculate its value.

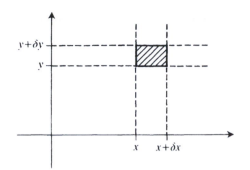

Fig. 6.1   The probability that (X, Y) lies in the shaded region is approximately $f_{X,Y}(x, y)\,\delta x\,\delta y$

Theorem 6A   *If A is any regular subset of $\mathbb{R}^2$ and X and Y are jointly continuous random variables with joint density function $f_{X,Y}$ then*

(17)
$$P((X, Y) \in A) = \iint\limits_{(x,y)\in A} f_{X,Y}(x, y)\,dx\,dy.$$

This is really a result about integration rather than about probability theory, and so we omit the proof; we do not even attempt to explain the term 'regular' applied to the set A, noting only that all 'nice' sets are regular (by definition of 'nice', actually), such as rectangles, discs, regions bounded by closed Jordan curves, and so on. On the other hand it is easy to see why (17) should hold. The set A may be split up into the union of lots of small non-overlapping rectangles and $f_{X,Y}(x, y)\,\delta x\,\delta y$ is the probability that (X, Y) takes a value in a typical rectangle; now, roughly speaking the probability that (X, Y) takes a value in A is the sum of these small probabilities.

Example 18   It is not too difficult to check that a function $f : \mathbb{R}^2 \to \mathbb{R}$ is the joint density function of some pair of random variables if and only if f satisfies (14) and (15):

$$f(x, y) \geq 0 \quad \text{for all } x \text{ and } y, \quad \text{and} \quad \int_{-\infty}^{\infty} \int_{-\infty}^{\infty} f(x, y)\,dx\,dy = 1.$$

This holds in just the same way as the corresponding properties (5.22) and (5.23) were necessary and sufficient for a function of one variable to be a density function. It follows that the function

$$f(x, y) = \begin{cases} 1/(ab) & \text{if } 0 < x < a \text{ and } 0 < y < b, \\ 0 & \text{otherwise,} \end{cases}$$

is a joint density function. If X and Y have joint density function f

then the vector (X, Y) is said to be *uniformly distributed* on the rectangle $B = (0, a) \times (0, b)$. If A is any region of the plane then

$$P((X, Y) \in A) = \iint_A f(x, y)\, dx\, dy$$

$$= \iint_{A \cap B} \frac{1}{ab}\, dx\, dy$$

$$= \frac{\text{area}\,(A \cap B)}{\text{area}\,(B)}. \qquad \square$$

Exercises 4. Random variables X and Y have joint density function

$$f(x, y) = \begin{cases} c(x^2 + \frac{1}{2}xy) & \text{if } 0 < x < 1,\ 0 < y < 2, \\ 0 & \text{otherwise.} \end{cases}$$

Find the value of the constant c and the joint distribution function of X and Y.

5. Random variables X and Y have joint density function

$$f(x, y) = \begin{cases} e^{-x-y} & \text{if } x, y > 0, \\ 0 & \text{otherwise.} \end{cases}$$

Find $P(X + Y \le 1)$ and $P(X > Y)$.

6.3 Marginal density functions and independence

Whenever X and Y have joint density function $f_{X,Y}$, the ordinary density functions of X and Y may be retrieved immediately since (at points of differentiability)

(19)
$$f_X(x) = \frac{d}{dx} P(X \le x)$$

$$= \frac{d}{dx} \int_{u=-\infty}^{x} \int_{v=-\infty}^{\infty} f_{X,Y}(u, v)\, du\, dv \qquad \text{by Theorem 6A}$$

$$= \int_{v=-\infty}^{\infty} f_{X,Y}(x, v)\, dv,$$

and similarly

(20)
$$f_Y(y) = \int_{u=-\infty}^{\infty} f_{X,Y}(u, y)\, du.$$

These density functions are called the *marginal* density functions of X

and of Y, since they are obtained by 'projecting' the random vector (X, Y) onto the two co-ordinate axes of the plane.

Recall that X and Y are called independent if their distribution functions satisfy

(21) $$F_{X,Y}(x, y) = F_X(x)F_Y(y) \qquad \text{for all } x, y \in \mathbb{R}.$$

If X and Y are jointly continuous, then differentiation of this relation with respect to both x and y yields a condition on their density functions,

(22) $$f_{X,Y}(x, y) = f_X(x)f_Y(y) \qquad \text{for all } x, y \in \mathbb{R},$$

and it is easy to see that (21) holds if and only if (22) holds—certainly (21) implies (22), and conversely if we integrate both sides of (22) as x ranges over $(-\infty, u]$ and y ranges over $(-\infty, v]$ then we obtain (21). Thus, jointly continuous random variables are independent if and only if their joint density function factorizes as the product of the two marginal density functions. This is exactly analogous to the case of discrete random variables, discussed in Section 3.3. Just as in the case of discrete random variables, there is a more general result.

Theorem 6B   *Jointly continuous random variables X and Y are independent if and only if their joint density function factorizes:*

(23) $$f_{X,Y}(x, y) = g(x)h(y) \qquad \text{for all } x, y \in \mathbb{R},$$

as the product of a function of the first variable and a function of the second.

We do not prove this, but suggest that the reader adapts the proof of Theorem 3B himself, replacing summations by integrals.

We do not wish to spend a lot of time going over the case when there are three or more random variables. Roughly speaking, all the ideas of this chapter so far have analogues in more than two dimensions. For example, three random variables X, Y, Z are called *jointly continuous* if

$$P(X \leq x, Y \leq y, Z \leq z) = \int_{u=-\infty}^{x} \int_{v=-\infty}^{y} \int_{w=-\infty}^{z} f(u, v, w) \, du \, dv \, dw$$

for all $x, y, z \in \mathbb{R}$ and some function f. If this holds then we may take

$$f(x, y, z) = \frac{\partial^3}{\partial x \, \partial y \, \partial z} P(X \leq x, Y \leq y, Z \leq z)$$

to be the joint density function of the triple (X, Y, Z), whenever these derivatives exist, and the random variables are independent if and only if f factorizes as the product of the marginal density functions:

$$f(x, y, z) = f_X(x)f_Y(y)f_Z(z) \qquad \text{for all } x, y, z \in \mathbb{R}.$$

Example 24   Suppose that X and Y have joint density function

$$f(x, y) = \begin{cases} e^{-x-y} & \text{if } x, y > 0, \\ 0 & \text{otherwise.} \end{cases}$$

Then

$$f_X(x) = \int_{-\infty}^{\infty} f(x, y) \, dy$$

$$= \begin{cases} \int_0^{\infty} e^{-x-y} \, dy & \text{if } x > 0, \\ 0 & \text{otherwise,} \end{cases}$$

$$= \begin{cases} e^{-x} & \text{if } x > 0, \\ 0 & \text{otherwise,} \end{cases}$$

giving that X has the exponential distribution with parameter 1. Y has this distribution also, and

$$f_{X,Y}(x, y) = f_X(x)f_Y(y) \qquad \text{for all } x, y \in \mathbb{R},$$

so that X and Y are independent. $\square$

Example 25   Suppose that X and Y have joint density function

(26)
$$f(x, y) = \begin{cases} ce^{-x-y} & \text{if } 0 < x < y, \\ 0 & \text{otherwise,} \end{cases}$$

for some constant c. Find c and ascertain whether or not X and Y are independent.

Solution   Joint density functions integrate to 1, so that

$$1 = \int_{-\infty}^{\infty} \int_{-\infty}^{\infty} f(x, y) \, dx \, dy$$

$$= c \int_{x=0}^{\infty} \int_{y=x}^{\infty} e^{-x-y} \, dx \, dy$$

$$= c \int_0^{\infty} e^{-2x} \, dx = \tfrac{1}{2}c,$$

giving that $c = 2$. Clearly X and Y are dependent, since f cannot be factorized as a function of x times a function of y for *all* pairs x, y

(look at the domain of f in (26)). More explicitly, by Theorem 6A,

$$P(X>2,\ Y<1) = \int_{x=2}^{\infty} \int_{y=-\infty}^{1} f(x,y)\, dx\, dy = 0$$

since $f(x,y)=0$ if $y<x$; on the other hand

$$P(X>2)>0 \quad \text{and} \quad P(Y<1)>0$$

so that

$$P(X>2,\ Y<1) \neq P(X>2)P(Y<1)$$

implying that X and Y cannot be independent. $\qquad\square$

Exercises 6. Let X and Y have joint density function

$$f(x,y) = \begin{cases} cx & \text{if } 0<y<x<1, \\ 0 & \text{otherwise.} \end{cases}$$

Find the value of the constant c and the marginal density functions of X and Y. Are X and Y independent?

7. Random variables X, Y, and Z have joint density function

$$f(x,y,z) = \begin{cases} 8xyz & \text{if } 0<x,y,z<1, \\ 0 & \text{otherwise.} \end{cases}$$

Are X, Y, and Z independent? Find $P(X>Y)$ and $P(Y>Z)$.

6.4 Sums of continuous random variables

We often need to know the density function of the sum $Z = X + Y$ of two jointly continuous random variables. The density function of Z is the derivative of the distribution function of Z, and so we calculate this first. Suppose that X and Y have joint density function $f_{X,Y}$. Then

$$P(Z \leq z) = P(X+Y \leq z)$$

$$= \iint_{A} f_{X,Y}(x,y)\, dx\, dy$$

by Theorem 6A, where $A = \{(x,y) \in \mathbb{R}^2 : x+y \leq z\}$. Writing in the limits of integration, we find that

$$P(Z \leq z) = \int_{x=-\infty}^{\infty} \int_{y=-\infty}^{z-x} f_{X,Y}(x,y)\, dx\, dy$$

$$= \int_{v=-\infty}^{z} \int_{u=-\infty}^{\infty} f_{X,Y}(u, v-u)\, du\, dv$$

by the substitution† $u = x$, $v = x + y$. Differentiate this equation with respect to z, where possible, to obtain

(27)
$$f_Z(z) = \int_{-\infty}^{\infty} f_{X,Y}(u, z - u) \, du.$$

An important special case is when X and Y are independent; in this case the following theorem is an immediate consequence of (27).

Theorem 6C

If the random variables X and Y are independent and jointly continuous with density functions f_X and f_Y, then the density function of $Z = X + Y$ is

(28)
$$f_Z(z) = \int_{-\infty}^{\infty} f_X(x) f_Y(z - x) \, dx \qquad \text{for } z \in \mathbb{R}.$$

In the language of analysis, equation (28) says that f_Z is the *convolution* of f_X and f_Y.

Example 29

Let X and Y be independent random variables having, respectively, the gamma distribution with parameters s and λ and the gamma distribution with parameters t and λ. Then $Z = X + Y$ has density function

$$f_Z(z) = \int_{-\infty}^{\infty} f_X(x) f_Y(z - x) \, dx$$

$$= \begin{cases} \int_0^z f_X(x) f_Y(z - x) \, dx & \text{if } z > 0, \\ 0 & \text{otherwise,} \end{cases}$$

since $f_X(x) f_Y(z - x) = 0$ unless $x > 0$ and $z - x > 0$. Thus, if $z > 0$ then

$$f_Z(z) = \int_0^z \frac{1}{\Gamma(s)} \lambda(\lambda x)^{s-1} e^{-\lambda x} \frac{1}{\Gamma(t)} \lambda[\lambda(z - x)]^{t-1} e^{-\lambda(z-x)} \, dx$$

$$= A e^{-\lambda z} \int_0^z x^{s-1} (z - x)^{t-1} \, dx$$

where

$$A = \frac{1}{\Gamma(s)\Gamma(t)} \lambda^{s+t}.$$

Substitute $y = x/z$ in the last integral to obtain

(30)
$$f_Z(z) = B z^{s+t-1} e^{-\lambda z} \qquad \text{if } z > 0,$$

† This is a simple change of variables in two dimensions. Those readers not familiar with such transformations in more than one dimension should read on to the next section.

where B is a constant given by

(31)
$$B = \frac{1}{\Gamma(s)\Gamma(t)} \lambda^{s+t} \int_0^1 y^{s-1}(1-y)^{t-1}\,dy.$$

The only distribution with density function having the form of (30) is the gamma distribution with parameters $s+t$ and λ, and it follows that the constant B is given by

(32)
$$B = \frac{1}{\Gamma(s+t)} \lambda^{s+t};$$

a glance at (5.28) confirms this. Our principal conclusion is that the sum of two independent gamma-distributed random variables, with parameters s, λ and t, λ respectively, has the gamma distribution with parameters $s+t$, λ. We have a subsidiary conclusion also; we compare (31) and (32) to find that

$$\int_0^1 y^{s-1}(1-y)^{t-1}\,dy = \frac{\Gamma(s)\Gamma(t)}{\Gamma(s+t)}$$

for any s, $t > 0$, a formula well known to experts at calculating integrals. □

Exercises 8. If X and Y have joint density function

$$f(x, y) = \begin{cases} \frac{1}{2}(x+y)e^{-x-y} & \text{if } x, y > 0, \\ 0 & \text{otherwise,} \end{cases}$$

find the density function of $X + Y$.
9. If X and Y are independent random variables having the χ^2 distribution with m and n degrees of freedom, respectively, prove that $X + Y$ has the χ^2 distribution with $m + n$ degrees of freedom.
10. If X and Y are independent random variables each having the normal distribution with mean 0 and variance 1, find the distribution of $X + Y$.

6.5 Changes of variables

The following type of question arises commonly: if X and Y are random variables and $u, v : \mathbb{R}^2 \to \mathbb{R}$, what can be said about the joint distribution of the pair (U, V) of random variables, given by $U = u(X, Y)$, $V = v(X, Y)$? We present an answer to this question in the particular case when X and Y are jointly continuous and u and v satisfy certain conditions which allow us to use the usual theory of changes of variables within an integral. Let T be the mapping from $\mathbb{R}^2$ into $\mathbb{R}^2$ given by $T(x, y) = (u, v)$ where $u = u(x, y)$ and $v = v(x, y)$, and suppose that T is a one–one mapping of some domain

$D \subseteq \mathbb{R}^2$ onto some range $S \subseteq \mathbb{R}^2$. Then T may be inverted to obtain a one-one onto mapping $T^{-1} : S \to D$; that is, for each $(u, v) \in S$, there exists a point $(x, y) = T^{-1}(u, v)$ in D, and we write $x = x(u, v)$ and $y = y(u, v)$. The *Jacobian* of T^{-1} is defined to be the determinant

$$(33) \qquad J(u, v) = \begin{vmatrix} \dfrac{\partial x}{\partial u} & \dfrac{\partial x}{\partial v} \\[2mm] \dfrac{\partial y}{\partial u} & \dfrac{\partial y}{\partial v} \end{vmatrix} = \frac{\partial x}{\partial u}\frac{\partial y}{\partial v} - \frac{\partial y}{\partial u}\frac{\partial x}{\partial v},$$

and we suppose that these derivatives exist and are continuous at all points in S. The standard theory of multiple integrals tells us how to change variables within the integral: if $g : \mathbb{R}^2 \to \mathbb{R}^2$ then, for any sufficiently regular subset A of D and decent function g,

$$(34) \qquad \iint_A g(x, y) \, dx \, dy = \iint_{T(A)} g(x(u, v), y(u, v)) \, |J(u, v)| \, du \, dv,$$

where $T(A)$ is the image of A under T.

Theorem 6D   *Let X and Y be jointly continuous with joint density function $f_{X,Y}$, and let $D = \{(x, y) : f_{X,Y}(x, y) > 0\}$. If the mapping T given by $T(x, y) = (u(x, y), v(x, y))$ maps D onto the set $S \subseteq \mathbb{R}^2$ and is invertible on S, then (subject to the previous conditions) the pair $(U, V) = (u(X, Y), v(X, Y))$ is jointly continuous with joint density function*

$$(35) \qquad f_{U,V}(u, v) = \begin{cases} f_{X,Y}(x(u, v), y(u, v)) \, |J(u, v)| & \text{if } (u, v) \in S, \\ 0 & \text{otherwise.} \end{cases}$$

Proof   You should not worry overmuch about the details of this argument. Suppose that $A \subseteq D$ and $T(A) = B$. The mapping T is one-one on D and hence

$$(36) \qquad P((U, V) \in B) = P((X, Y) \in A).$$

However,

$$P((X, Y) \in A) = \iint_A f_{X,Y}(x, y) \, dx \, dy \qquad \text{by Theorem } 6A$$

$$= \iint_B f_{X,Y}(x(u, v), y(u, v)) \, |J(u, v)| \, du \, dv \qquad \text{by (34)}$$

$$= P((U, V) \in B) \qquad \text{by (36).}$$

This holds for any $B \subseteq S$, and another glance at Theorem $6A$ gives the result. $\square$

Although the statement of Theorem 6D looks forbidding, it is not difficult to apply in practice, although it is necessary to check that the mapping in question is one-one. Here is an example.

Example 37

Let X and Y have joint density function

$$f(x, y) = \begin{cases} e^{-x-y} & \text{if } x, y > 0, \\ 0 & \text{otherwise,} \end{cases}$$

and let $U = X + Y$ and $V = X/(X + Y)$. Find the joint density function of U and V and the marginal density function of V.

Solution

The mapping T of this problem is given by $T(x, y) = (u, v)$ where

$$u = x + y, \qquad v = \frac{x}{x + y},$$

and T maps $D = \{(x, y) : x, y > 0\}$ onto $S = \{(u, v) : 0 < u < \infty, \ 0 < v < 1\}$ with inverse $T^{-1}(u, v) = (x, y)$ where

$$x = uv, \qquad y = u(1 - v).$$

The Jacobian of T^{-1} is

$$\begin{vmatrix} \dfrac{\partial x}{\partial u} & \dfrac{\partial x}{\partial v} \\[2mm] \dfrac{\partial y}{\partial u} & \dfrac{\partial y}{\partial v} \end{vmatrix} = \begin{vmatrix} v & u \\ (1 - v) & -u \end{vmatrix}$$

$$= -u,$$

giving from (35) that U and V have joint density function

$$f_{U,V}(u, v) = \begin{cases} ue^{-u} & \text{if } u > 0 \text{ and } 0 < v < 1, \\ 0 & \text{otherwise.} \end{cases}$$

The marginal density function of V is

$$f_V(v) = \int_{-\infty}^{\infty} f_{U,V}(u, v)\, du$$

$$= \begin{cases} \displaystyle\int_0^{\infty} ue^{-u}\, du = 1 & \text{if } 0 < v < 1, \\[2mm] 0 & \text{otherwise,} \end{cases}$$

so that V is uniformly distributed on $(0, 1)$. Actually, U and V are independent and U has the gamma distribution with parameters 2 and 1. □

Exercises

11. Let X and Y be independent random variables each having the normal distribution with mean μ and variance σ^2. Find the joint density function of $U = X - Y$ and $V = X + Y$. Are U and V independent?

12. Let X and Y be random variables with joint density function

$$f(x, y) = \begin{cases} \frac{1}{4}e^{-\frac{1}{2}(x+y)} & \text{if } x, y > 0, \\ 0 & \text{otherwise.} \end{cases}$$

Show that the joint density function of $U = \frac{1}{2}(X - Y)$ and $V = Y$ is

$$f_{U,V}(u, v) = \begin{cases} \frac{1}{2}e^{-u-v} & \text{if } (u, v) \in A, \\ 0 & \text{otherwise,} \end{cases}$$

where A is a region of the (u, v) plane to be determined. Deduce that U has the bilateral exponential distribution with density function

$$f_U(u) = \frac{1}{2}e^{-|u|} \qquad \text{for } u \in \mathbb{R}.$$

6.6 Conditional density functions

Let us suppose that X and Y are jointly continuous random variables with joint density function $f_{X,Y}$. To obtain the marginal density function f_Y of Y, we 'average' over all possible values of X,

$$f_Y(y) = \int_{-\infty}^{\infty} f_{X,Y}(x, y) \, dx,$$

and this is the calculation which we perform if we care about Y only and have no information about the value taken by X. A starkly contrasting situation arises if we have full information about the value taken by X, say we are given that X takes the value x; this information has consequences for the distribution of Y, and it is this 'conditional' distribution of Y given that $X = x$' that concerns us in this section. We cannot calculate $P(Y \le y \mid X = x)$ from the usual formula $P(A \mid B) = P(A \cap B)/P(B)$ since $P(B) = 0$ in this case, and so we proceed as follows. Instead of conditioning on the event that $X = x$, we condition on the event that $x \le X \le x + \delta x$ and take a limit as $\delta x \downarrow 0$. Thus

$$P(Y \le y \mid x \le X \le x + \delta x) = \frac{P(Y \le y, \, x \le X \le x + \delta x)}{P(x \le X \le x + \delta x)}$$

$$= \frac{\displaystyle\int_{u=x}^{x+\delta x} \int_{v=-\infty}^{y} f_{X,Y}(u, v) \, du \, dv}{\displaystyle\int_{x}^{x+\delta x} f_X(u) \, du}$$

by Theorems 6A and 5A. We divide both the numerator and the

denominator by δx and take the limit as $\delta x \downarrow 0$ to obtain

(38)
$$P(Y \leq y \mid x \leq X \leq x + \delta x) \to \int_{-\infty}^{y} \frac{f_{X,Y}(x, v)}{f_X(x)} dv$$
$$= G(y),$$

say. It is clear from (38) that G is a distribution function with density function

$$g(y) = \frac{f_{X,Y}(x, y)}{f_X(x)} \qquad \text{for } y \in \mathbb{R},$$

and we call G and g the 'conditional distribution function' and the 'conditional density function' of Y given that X equals x. The above discussion is valid only for values of x such that $f_X(x) > 0$, and so we make the following formal definition. The *conditional density function of Y given that $X = x$* is denoted by $f_{Y|X}(\bullet \mid x)$ and defined by

(39)
$$f_{Y|X}(y \mid x) = \frac{f_{X,Y}(x, y)}{f_X(x)}$$

for all $y \in \mathbb{R}$ and all x such that $f_X(x) > 0$.

We emphasize that expressions such as $P(Y \leq y \mid X = x)$ cannot be interpreted in the usual way by using the formula for $P(A \mid B)$. The only way of giving meaning to such a quantity is to make a new definition, such as: $P(Y \leq y \mid X = x)$ is defined to be the conditional distribution function $G(y)$ of Y given $X = x$, given in (38).

If X and Y are independent then $f_{X,Y}(x, y) = f_X(x)f_Y(y)$ and (39) gives that $f_{Y|X}(y \mid x) = f_Y(y)$, which is to say that information about X is irrelevant when studying Y.

Example 40   Let X and Y have joint density function

$$f(x, y) = \begin{cases} 2e^{-x-y} & \text{if } 0 < x < y < \infty, \\ 0 & \text{otherwise.} \end{cases}$$

The marginal density functions are

$$f_X(x) = 2e^{-2x} \qquad \text{for } x > 0,$$

and

$$f_Y(y) = 2e^{-y}(1 - e^{-y}) \qquad \text{for } y > 0,$$

where it is understood that these functions take the value 0 off the specified domains of x and y. The conditional density function of Y given $X = x$ (> 0) is

$$f_{Y|X}(y \mid x) = \frac{2e^{-x-y}}{2e^{-2x}}$$
$$= e^{x-y} \qquad \text{for } y > x.$$

The conditional density function of X given $Y = y$ is

$$f_{X|Y}(x \mid y) = \frac{e^{-x}}{1 - e^{-y}} \qquad \text{if } 0 < x < y.$$

It is clear that both these conditional density functions equal 0 if $x > y$. $\qquad\square$

Exercises   13. Suppose that X and Y have joint density function

$$f(x, y) = \begin{cases} e^{-y} & \text{if } 0 < x < y < \infty, \\ 0 & \text{otherwise.} \end{cases}$$

Find the conditional density functions of X given that $Y = y$ and of Y given that $X = x$.

14. Let X and Y be independent random variables each having the exponential distribution with parameter λ. Find the joint density function of X and $X + Y$, and deduce that the conditional density function of X given that $X + Y = a$ is uniform on the interval $(0, a)$ for each $a > 0$. In other words, the knowledge that $X + Y = a$ provides no useful clue about the position of X in the interval $(0, a)$.

6.7   Expectations of continuous random variables

Let X and Y be jointly continuous random variables on $(\Omega, \mathcal{F}, P)$, and let $g : \mathbb{R}^2 \to \mathbb{R}$. As in the discussion in Section 5.5, we shall suppose that the mapping $Z : \Omega \to \mathbb{R}$, defined by $Z(\omega) = g(X(\omega), Y(\omega))$, is a random variable (this is certainly the case if g is sufficiently well behaved). As before, in calculating the expectation of Z, we do not have to find the distribution of Z explicitly.

Theorem 6E

$$E(g(X, Y)) = \int_{-\infty}^{\infty} \int_{-\infty}^{\infty} g(x, y) f_{X,Y}(x, y) \, dx \, dy,$$

whenever this integral converges absolutely.

We do not prove this, but note that the result follows intuitively from the corresponding result (Theorem 3A) for discrete random variables, by exploiting the analogy between joint mass functions and joint density functions.

Using Theorem 6E, we find that the expectation operator acts linearly on the space of continuous random variables. This may be restated in rather less grandiose language by saying that

(41)
$$E(aX + bY) = aE(X) + bE(Y)$$

whenever $a, b \in \mathbb{R}$ and X and Y are jointly continuous random variables with means $E(X)$ and $E(Y)$. This follows from Theorem 6E by writing

$$E(aX + bY) = \int_{-\infty}^{\infty} \int_{-\infty}^{\infty} (ax + by) f_{X,Y}(x, y) \, dx \, dy$$

$$= a \int_{-\infty}^{\infty} \int_{-\infty}^{\infty} x f_{X,Y}(x, y) \, dx \, dy + b \int_{-\infty}^{\infty} \int_{-\infty}^{\infty} y f_{X,Y}(x, y) \, dx \, dy$$

$$= a \int_{-\infty}^{\infty} x f_X(x) \, dx + b \int_{-\infty}^{\infty} y f_Y(y) \, dy$$

$$= aE(X) + bE(Y).$$

We mention one common error here. It is a mistake to demand that X and Y be independent in order that $E(X + Y) = E(X) + E(Y)$; this equation holds *whether or not* X and Y are independent.

Next we discuss the relationship between expectation and independence, noting the excellent analogy with Theorems 3C and 3D which dealt with discrete random variables. First, note that if X and Y are independent random variables with joint density function $f_{X,Y}$ then

(42) $$E(XY) = E(X)E(Y)$$

whenever these expectations exist, since

(43) $$E(XY) = \int_{-\infty}^{\infty} \int_{-\infty}^{\infty} xy f_{X,Y}(x, y) \, dx \, dy$$

$$= \int_{-\infty}^{\infty} x f_X(x) \, dx \int_{-\infty}^{\infty} y f_Y(y) \, dy \qquad \text{by independence}$$

$$= E(X)E(Y).$$

The converse is false: there exist jointly continuous dependent random variables X and Y for which $E(XY) = E(X)E(Y)$. The correct and full result here is the next theorem.

Theorem 6F *Jointly continuous random variables X and Y are independent if and only if*

(44) $$E(g(X)h(Y)) = E(g(X))E(h(Y))$$

for all functions $g, h : \mathbb{R} \to \mathbb{R}$ for which these expectations exist.

Proof If X and Y are independent then $g(X)$ and $h(Y)$ are independent (this is a little technical to prove, but intuitively clear), and hence (44) holds by applying (42) to $g(X)$ and $h(Y)$. Conversely, if (44) holds for all appropriate functions g and h then it holds in particular for the functions given by

$$g(u) = \begin{cases} 1 & \text{if } u \leq x, \\ 0 & \text{if } u > x, \end{cases}$$

$$h(v) = \begin{cases} 1 & \text{if } v \leq y, \\ 0 & \text{if } v > y, \end{cases}$$

for fixed values of x and y. In this case $g(X)h(Y)$ is a discrete random variable with the Bernoulli distribution, parameter $p_1 = P(X \leq x, Y \leq y)$, and $g(X)$ and $h(Y)$ are Bernoulli random variables with parameters $p_2 = P(X \leq x)$ and $p_3 = P(Y \leq y)$ respectively. Hence

$$E(g(X)h(Y)) = P(X \leq x, Y \leq y)$$

by (2.19), and

$$E(g(X)) = P(X \leq x), \quad E(h(Y)) = P(Y \leq y),$$

giving by (44) that

$$P(X \leq x, Y \leq y) = P(X \leq x)P(Y \leq y) \qquad \text{for } x, y \in \mathbb{R},$$

as required. $\square$

Exercises 15. Let the pair (X, Y) be uniformly distributed on the unit disc, so that

$$f_{X,Y}(x, y) = \begin{cases} \pi^{-1} & \text{if } x^2 + y^2 \leq 1, \\ 0 & \text{otherwise.} \end{cases}$$

Find $E\sqrt{(X^2 + Y^2)}$ and $E(X^2 + Y^2)$.
16. Give an example of a pair of dependent and jointly continuous random variables X, Y for which

$$E(XY) = E(X)E(Y).$$

6.8 Conditional expectation and the bivariate normal distribution

Suppose that X and Y are jointly continuous random variables with joint density function $f_{X,Y}$, and that we are given that $X = x$. In the light of this information, the new density function of Y is the conditional density function $f_{Y|X}(\bullet \mid x)$. The mean of this density function is called the *conditional expectation of Y given $X = x$* and denoted by $E(Y \mid X = x)$:

(45)
$$E(Y \mid X = x) = \int_{-\infty}^{\infty} y f_{Y|X}(y \mid x) \, dy$$

$$= \int_{-\infty}^{\infty} y \frac{f_{X,Y}(x, y)}{f_X(x)} \, dy,$$

defined for any value of x for which $f_X(x) > 0$.

Possibly the most useful application of conditional expectations is in the next theorem, a form of the partition theorem which enables us to calculate $E(Y)$ in situations where the conditional expectations $E(Y \mid X = x)$ are easily calculated.

Theorem 6G

If X and Y are jointly continuous random variables then

(46)
$$E(Y) = \int E(Y \mid X = x) f_X(x) \, dx$$

where the integral is over all values of x such that $f_X(x) > 0$.

In other words, in calculating $E(Y)$ we may first fix the value of X and then average over this value later.

Proof

This is very straightforward:

$$E(Y) = \int y f_Y(y) \, dy$$

$$= \int\int y f_{X,Y}(x, y) \, dx \, dy$$

$$= \int\int y f_{Y|X}(y \mid x) f_X(x) \, dx \, dy \qquad \text{by (39)}$$

$$= \int \left(\int y f_{Y|X}(y \mid x) \, dy \right) f_X(x) \, dx$$

as required. The integrals here range over all appropriate values of x and y.   □

Example 47

Bivariate normal distribution. This is a good example of many of the ideas of this chapter. Let ρ be a number satisfying $-1 < \rho < 1$, and let f be the function of two variables given by

(48)
$$f(x, y) = \frac{1}{2\pi\sqrt{(1 - \rho^2)}} \exp\left(-\frac{1}{2(1 - \rho^2)} (x^2 - 2\rho xy + y^2)\right) \quad \text{for } x, y \in \mathbb{R}.$$

Clearly $f(x, y) \geq 0$ for all x and y, and it is the case that $\int_{-\infty}^{\infty} \int_{-\infty}^{\infty} f(x, y) \, dx \, dy = 1$ (the reader should check this), giving that f is a joint density function; it is called the joint density function of the *standard bivariate normal distribution*. Suppose that X and Y are random variables with the standard bivariate normal density function f. In this example, we shall calculate
  (i) the marginal density function of X,
  (ii) the conditional density function of Y given $X = x$,
  (iii) the conditional expectation of Y given $X = x$,
  (iv) a condition for X and Y to be independent.

Marginals. The marginal density function of X is

$$f_X(x) = \int_{-\infty}^{\infty} f(x, y)\, dy$$

$$= \int_{-\infty}^{\infty} \frac{1}{2\pi\sqrt{(1-\rho^2)}} \exp\left(-\frac{1}{2(1-\rho^2)}[(y-\rho x)^2 + x^2(1-\rho^2)]\right) dy$$

$$= \frac{1}{\sqrt{(2\pi)}} e^{-\frac{1}{2}x^2} \int_{-\infty}^{\infty} \frac{1}{\sqrt{[2\pi(1-\rho^2)]}} \exp\left(-\frac{(y-\rho x)^2}{2(1-\rho^2)}\right) dy.$$

The function within the integral here is the density function of the normal distribution with mean ρx and variance $1-\rho^2$, and therefore this final integral equals 1, giving that

(49)
$$f_X(x) = \frac{1}{\sqrt{(2\pi)}} e^{-\frac{1}{2}x^2} \qquad \text{for } x \in \mathbb{R},$$

so that X has the normal distribution with mean 0 and variance 1. By symmetry, Y has this distribution also.

Conditional density function. The conditional density function of Y given that $X = x$ is

$$f_{Y|X}(y \mid x) = \frac{f(x, y)}{f_X(x)}$$

$$= \frac{1}{\sqrt{[2\pi(1-\rho^2)]}} \exp\left(-\frac{(y-\rho x)^2}{2(1-\rho^2)}\right),$$

and so the conditional distribution of Y given $X = x$ is the normal distribution with mean ρx and variance $1-\rho^2$.

Conditional expectation. The conditional expectation of Y given $X = x$ is

(50)
$$E(Y \mid X = x) = \int_{-\infty}^{\infty} y f_{Y|X}(y \mid x)\, dy$$

$$= \rho x,$$

from the above remarks.

Independence. The random variables X and Y are independent if and only if $f(x, y)$ factorizes as the product of a function of x and a function of y, and this happens (by a glance at (48)) if and only if $\rho = 0$. The constant ρ occurs in another way also. We may calculate $E(XY)$ by applying Theorem 6G to the random variables X and XY

to obtain

$$E(XY) = \int_{-\infty}^{\infty} E(XY \mid X = x) f_X(x) \, dx$$

$$= \int_{-\infty}^{\infty} x E(Y \mid X = x) f_X(x) \, dx$$

$$= \int_{-\infty}^{\infty} \rho x^2 f_X(x) \, dx \qquad \text{by (50)}$$

$$= \rho E(X^2)$$

$$= \rho \, \text{var}(X) \qquad \text{since } E(X) = 0$$

$$= \rho$$

by the remarks after (49). Also $E(X) = E(Y) = 0$, giving that

$$\rho = E(XY) - E(X)E(Y).$$

We deduce that X and Y are independent if and only if $E(XY) = E(X)E(Y)$; thus, by the discussion around (43), for random variables X and Y with the bivariate normal distribution, *X and Y are independent if and only if* $E(XY) = E(X)E(Y)$.

Finally, we note a more general bivariate normal distribution. Let g be the function of two variables given by

(51) $$g(x, y) = \frac{1}{2\pi\sigma_1\sigma_2\sqrt{(1 - \rho^2)}} \exp(-\tfrac{1}{2}Q(x, y)) \qquad \text{for } x, y \in \mathbb{R},$$

where

(52) $$Q(x, y) = \frac{1}{1 - \rho^2} \left[\left(\frac{x - \mu_1}{\sigma_1} \right)^2 - 2\rho \left(\frac{x - \mu_1}{\sigma_1} \right) \left(\frac{y - \mu_2}{\sigma_2} \right) + \left(\frac{y - \mu_2}{\sigma_2} \right)^2 \right]$$

and $\mu_1, \mu_2 \in \mathbb{R}$, $\sigma_1, \sigma_2 > 0$, $-1 < \rho < 1$. The standard bivariate normal distribution is obtained by setting $\mu_1 = \mu_2 = 0$, $\sigma_1 = \sigma_2 = 1$. It is not difficult but slightly tedious to show that g is a joint density function, and the corresponding distribution is called the *bivariate normal distribution* with the appropriate parameters. We leave it as an exercise to show that if U and V have joint density function g then the pair X, Y given by

(53) $$X = \frac{U - \mu_1}{\sigma_1}, \qquad Y = \frac{V - \mu_2}{\sigma_2}$$

has the standard bivariate normal distribution with parameter ρ. □

Exercises   17. If X and Y have joint density function

$$f(x, y) = \begin{cases} e^{-y} & \text{if } 0 < x < y < \infty, \\ 0 & \text{otherwise,} \end{cases}$$

find $E(X \mid Y = y)$ and $E(Y \mid X = x)$.
18. Verify the assertion above: if the pair (U, V) has the bivariate normal density function (51) and X and Y are given by (53), then X and Y have the standard bivariate normal distribution. Hence or otherwise show that

$$E(UV) - E(U)E(V) = \rho\sigma_1\sigma_2,$$

and that

$$E(V \mid U = u) = \mu_2 + \rho\sigma_2(u - \mu_1)/\sigma_1.$$

6.9   Problems

1. If X and Y are independent random variables with density functions f_X and f_Y respectively, show that $U = XY$ and $V = X/Y$ have density functions

$$f_U(u) = \int_{-\infty}^{\infty} f_X(x) f_Y(u/x) |x|^{-1} \, dx, \qquad f_V(v) = \int_{-\infty}^{\infty} f_X(vy) f_Y(y) |y| \, dy.$$

2. Is the function G, defined by

$$G(x, y) = \begin{cases} 1 & \text{if } x + y \geq 0, \\ 0 & \text{otherwise,} \end{cases}$$

the joint distribution function of some pair of random variables? Justify your answer.
3. Let (X, Y, Z) be a point chosen uniformly at random in the unit cube $(0, 1) \times (0, 1) \times (0, 1)$. Find the probability that the quadratic function of t,

$$Xt^2 + Yt + Z = 0,$$

has two distinct real roots.
4. Show that if X and Y are independent random variables having the exponential distribution with parameters λ and μ respectively, then $\min\{X, Y\}$ has the exponential distribution with parameter $\lambda + \mu$.
5. If X has the exponential distribution, show that

$$P(X > u + v \mid X > u) = P(X > v) \qquad \text{for all } u, v > 0.$$

This is called the 'lack-of-memory' property, since it says that, if we are given that $X > u$, then the distribution of $X - u$ is the same as the original distribution of X. Show that if Y is a positive random variable with the lack-of-memory property above, then Y has the exponential distribution.
6. Let $X_1, X_2, \ldots, X_n$ be independent random variables, each having distribution function F and density function f. Find the distribution function of U and the density functions of U and V, where $U =$

$\min\{X_1, X_2, \ldots, X_n\}$ and $V = \max\{X_1, X_2, \ldots, X_n\}$. Show that the joint density function of U and V is

$$f_{U,V}(u, v) = n(n-1)f(u)f(v)[F(v) - F(u)]^{n-2} \quad \text{if} \quad u < v.$$

7. Let $X_1, X_2, \ldots$ be independent identically distributed continuous random variables. Define N as the unique index such that

$$X_1 \geq X_2 \geq \cdots \geq X_{N-1} \quad \text{and} \quad X_{N-1} < X_N.$$

Prove that $P(N = k) = (k-1)/k!$ and that $E(N) = e$.

8. Show that there exists a constant c such that the function

$$f(x, y) = \frac{c}{(1 + x^2 + y^2)^{\frac{3}{2}}} \qquad \text{for } x, y \in \mathbb{R}$$

is a joint density function. Show that both marginal density functions of f are the density function of the Cauchy distribution.

9. Suppose random variables X and Y have joint distribution function $F_{X,Y}$ and joint density function $f_{X,Y}$ given by

$$f_{X,Y}(x, y) = \begin{cases} \frac{1}{3}(x + 2y)e^{-(x+y)} & \text{if } x, y \geq 0, \\ 0 & \text{otherwise.} \end{cases}$$

Find the marginal density function for Y and the conditional density function for X given that $Y = y$ (where $y > 0$). Show that $P(Y > X) = \frac{7}{12}$. Let $W = \min\{X, Y\}$, $Z = \max\{X, Y\}$, and let $F_{W,Z}$ denote the joint distribution function for W and Z. Let real numbers w and z be given with $0 < w < z$. Find an expression for $F_{W,Z}(w, z)$ in terms of $F_{X,Y}(w, z)$, $F_{X,Y}(z, w)$, and $F_{X,Y}(w, w)$ and hence find the joint density function for W and Z.
(Bristol 1983)

10. A particle is emitted from a source at time zero. During flight it undergoes a decay at some random time S and one of the decay products is then observed at time T, where $T \geq S$. Suppose T has a gamma distribution with density

$$f_T(t) = \begin{cases} \lambda^2 t e^{-\lambda t} & \text{if } t > 0, \\ 0 & \text{if } t \leq 0; \end{cases}$$

and suppose that for each $t > 0$ the conditional distribution of S given $T = t$ is uniform over the interval $(0, t)$. Find the joint density function of S and T. Hence find the joint density function of S and $T - S$ and show that these random variables are independent and identically distributed. Let $Z = \max\{S, T - S\}$; find the density function of Z.
(Bristol 1980)

11. Let $X_1, X_2, \ldots$ be a sequence of independent, identically distributed random variables, each having an exponential distribution with parameter λ. Define $S_0 = 0$, and for each positive integer n define

$$S_n = X_1 + \cdots + X_n.$$

Use mathematical induction to show that S_n has density

$$f_n(x) = \begin{cases} \dfrac{\lambda^n}{(n-1)!} x^{n-1}e^{-\lambda x} & \text{if } x > 0, \\ 0 & \text{if } x \leq 0, \end{cases}$$

for all $n \geq 1$. Now let $t > 0$ be given. Let $N(t) = \max\{n : S_n \leq t\}$. Show that $N(t)$ has a Poisson distribution. (Bristol 1980)

12. A random variable X has probability density function $f_X(x)$; the measurement of X is subject to a certain error, so that an attempt to record X results in the recording of a quantity $Z = X + Y$, where Y is a random variable independent of X having probability density function $f_Y(y)$. Show that the probability density function $f_Z(z)$ of Z is given by

$$f_Z(z) = \int_{-\infty}^{\infty} f_Y(u) f_X(z - u) \, du.$$

If

$$f_X(x) = \begin{cases} x \exp(-\tfrac{1}{2}x^2) & \text{if } x \geq 0, \\ 0 & \text{otherwise,} \end{cases}$$

and Y has a uniform distribution on $[-\varepsilon, \varepsilon]$, find $f_Z(z)$. Show that

$$P(Z > \varepsilon) = \frac{1}{2\varepsilon} \int_0^{2\varepsilon} e^{-\frac{1}{2}u^2} \, du. \qquad \text{(Bristol 1981)}$$

13. If X and Y are independent $N(0, \sigma^2)$ random variables show that $Z = Y/X$ has the Cauchy density function. Deduce that if (R, θ) is the representation in polar coordinates of the point (X, Y) in the Cartesian plane then θ is uniformly distributed on $[0, 2\pi]$. What is the distribution of R^2? (Bristol 1978)

14. An aeroplane is dropping medical supplies to two duellists. With respect to Cartesian coordinates whose origin is at the target point, both the x and y coordinates of the landing point of the supplies have normal distributions which are independent. These two distributions have the same mean 0 and variance σ^2. Show that the expectation of the distance between the landing point and the target is $\sigma\sqrt{(\tfrac{1}{2}\pi)}$. What is the variance of this distance? (Oxford 1976M)

15. X and Y are independent random variables normally distributed with mean zero and variance σ^2. Find the expectation of $(X^2 + Y^2)^{\frac{1}{2}}$. Find the probabilities of the following events, where a, b, c, and α are positive constants such that $b < c$ and $\alpha < \tfrac{1}{2}\pi$:
   (i) $(X^2 + Y^2)^{\frac{1}{2}} < a$;
   (ii) $0 < \arctan(Y/X) < \alpha$ and $Y > 0$;
   (iii) $b < (X^2 + Y^2)^{\frac{1}{2}} < c$ and $\tfrac{1}{4}\pi < \arctan(Y/X) < \tfrac{1}{2}\pi$, given that $(X^2 + Y^2)^{\frac{1}{2}} < a$, $0 < \arctan(Y/X) < \tfrac{1}{3}\pi$ and $Y > 0$.
   (Consider various cases depending on the relative sizes of a, b, and c.)
   (Oxford 1981M)

16. The independent random variables X and Y are both exponentially distributed with parameter λ, that is, each has density function

$$f(t) = \begin{cases} \lambda e^{-\lambda t} & \text{if } t > 0, \\ 0 & \text{if } t \leq 0. \end{cases}$$

   (a) Find the (cumulative) distribution and density functions of the random variables $1 - e^{-\lambda X}$, $\min\{X, Y\}$ and $X - Y$.
   (b) Find the probability that $\max\{X, Y\} \leq aX$ where a is a real constant. (Oxford 1982M)

17. The independent random variables X and Y are normally distributed with mean 0 and variance 1.
    (i) Show that $W = 2X - Y$ is normally distributed, and find its mean and variance.
    (ii) Find the mean of $Z = X^2/(X^2 + Y^2)$.
    (iii) Find the mean of V/U where $U = \max\{|X|, |Y|\}$ and $V = \min\{|X|, |Y|\}$. (Oxford 1985M)

    The two distributions which follow are important in statistical theory.
18. Let X and Y be independent random variables, X having the normal distribution with mean 0 and variance 1, and Y having the χ^2 distribution with n degrees of freedom. Show that

$$T = \frac{X}{\sqrt{(Y/n)}}$$

has density function

$$f(t) = \frac{1}{\sqrt{(\pi n)}} \frac{\Gamma(\frac{1}{2}(n+1))}{\Gamma(\frac{1}{2}n)} \left(1 + \frac{t^2}{n}\right)^{-\frac{1}{2}(n+1)} \quad \text{for } t \in \mathbb{R}.$$

T is said to have the *t-distribution* with n degrees of freedom.
19. Let X and Y be independent random variables with the χ^2 distribution, X having m degrees of freedom and Y having n degrees of freedom. Show that

$$U = \frac{X/m}{Y/n}$$

has density function

$$f(u) = \frac{m\Gamma(\frac{1}{2}(m+n))}{n\Gamma(\frac{1}{2}m)\Gamma(\frac{1}{2}n)} \frac{(mu/n)^{\frac{1}{2}m-1}}{[1 + (mu/n)]^{\frac{1}{2}(m+n)}} \quad \text{for } u > 0.$$

U is said to have the *F-distribution* with m and n degrees of freedom.
20. In a sequence of dependent Bernoulli trials the conditional probability of success at the ith trial, given that all preceding trials have resulted in failure, is p_i ($i = 1, 2, \ldots$). Give an expression in terms of the p_i for the probability that the first success occurs at the nth trial. Suppose that $p_i = 1/(i + 1)$ and that the time intervals between successive trials are independent random variables, the interval between the $(n - 1)$th and the nth trials being exponentially distributed with density $n^\alpha \exp(-n^\alpha x)$ where α is a given constant. Show that the expected time to achieve the first success is finite if and only if $\alpha > 0$. (Oxford 1975F)
21. Independent positive random variables X and Y have probability densities

$$x^{a-1}e^{-x}/\Gamma(a), \quad y^{b-1}e^{-y}/\Gamma(b) \quad (x, y \geq 0; a, b > 0),$$

respectively, and U and V are defined by

$$U = X + Y, \quad V = \frac{X}{X + Y}.$$

Prove that U and V are independent, and find their distributions.

Deduce that

$$E\left(\frac{X}{X+Y}\right) = \frac{E(X)}{E(X)+E(Y)}.$$ (Oxford 1971F)

22. Let X_1, X_2, X_3 be independent χ^2 random variables with r_1, r_2, r_3 degrees of freedom.
    (i) Show that $Y_1 = X_1/X_2$ and $Y_2 = X_1 + X_2$ are independent and that Y_2 is a χ^2 random variable with $r_1 + r_2$ degrees of freedom.
    (ii) Deduce that the following random variables are independent:

$$\frac{X_1/r_1}{X_2/r_2} \quad \text{and} \quad \frac{X_3/r_3}{(X_1+X_2)/(r_1+r_2)}.$$ (Oxford 1982F)

7

Moments, and moment generating functions

7.1 A general note

Up to now we have treated discrete and continuous random variables separately, and have hardly broached the existence of random variables which are neither discrete nor continuous. A brief overview of the material so far might look like this:

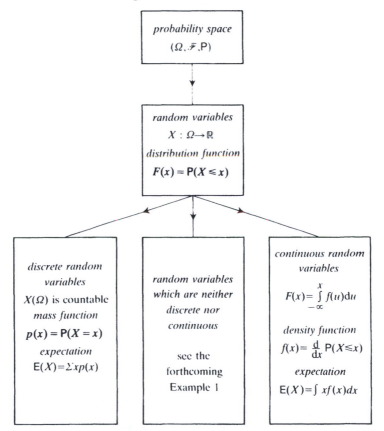

It is clear that we cannot carry on in this way forever, treating the discrete case and the continuous case separately and the others not at all. The correct thing to be done at this point is to study random variables in their entirety. Unfortunately, such a proper treatment is too advanced for this primitive text since it involves defining the expectation of an arbitrary random variable, using deep ideas of abstract measure and integration theory. We are therefore forced to adopt another strategy. We shall try to state and prove theorems in ways which do not explicitly mention the type (discrete or continuous or ...) of the random variables involved; generally speaking such arguments may be assumed to hold in the wide sense. Sometimes we will have to consider special cases, and then we shall normally treat *continuous* random variables; the discrete case is usually similar and easier, and a rule of thumb for converting an argument about continuous random variables into an argument about discrete random variables is to replace $f_X(x)\,dx$ by $p_X(x)$ and $\int$ by $\sum$.

Finally, curious readers may care to see the standard example of a random variable whose type is neither discrete nor continuous; less enthusiastic readers should go directly to the next section.

Example 1

The Cantor distribution. The celebrated Cantor set C is the often quoted example of an uncountable subset of the real line which is very sparse, in the sense that for any $\varepsilon > 0$ there exist intervals I_1, $I_2, \ldots$ with total length less than ε such that $C \subseteq \bigcup_n I_n$. We construct this set as follows. Let $C_1 = [0, 1]$. Delete the middle third $(\frac{1}{3}, \frac{2}{3})$ of C_1 and let $C_2 = [0, \frac{1}{3}] \cup [\frac{2}{3}, 1]$ be the remaining set. Next, delete the middle third in each of the two intervals comprising C_2 to obtain $C_3 = [0, \frac{1}{9}] \cup [\frac{2}{9}, \frac{1}{3}] \cup [\frac{2}{3}, \frac{7}{9}] \cup [\frac{8}{9}, 1]$, and continue similarly to obtain an infinite nested sequence $C_1 \supseteq C_2 \supseteq C_3 \supseteq \cdots$ of subsets of $[0, 1]$. The Cantor set C is defined to be the limit

$$C = \lim_{n \to \infty} C_n = \bigcap_{i=1}^{\infty} C_i$$

of this sequence. There is another way of thinking about the Cantor set and this is useful for us. Just as each number in $[0, 1]$ has an expansion in the base-10 system (namely its decimal expansion) so it has an expansion in the base-3 system. That is to say, any $x \in [0, 1]$ may be written in the form

$$x = \sum_{i=1}^{\infty} \frac{a_i}{3^i}$$

where each of the a's equals 0, 1, or 2. The Cantor set C is the set of all points $x \in [0, 1]$ for which the a's above take the values 0 and 2 only.

We obtain the *Cantor distribution* as follows. Take $x \in C$ and express x in the form described above:

$$x = \sum_{i=1}^{\infty} \frac{a_i}{3^i}$$

where each a_i equals 0 or 2. We define $F(x)$ by

$$F(x) = \sum_{i=1}^{\infty} \frac{b_i}{2^i}$$

where $b_i = \frac{1}{2}a_i$, so that

$$b_i = \begin{cases} 0 & \text{if } a_i = 0, \\ 1 & \text{if } a_i = 2. \end{cases}$$

It is clear that

$$F(0) = 0 \quad \text{and} \quad F(1) = 1,$$

and F is non-decreasing in that

$$F(x) \leq F(y) \qquad \text{if } x \leq y.$$

However, F is not a distribution function since it is defined on C only, but we may extend the domain of F to the whole real line in a natural way. If $x \in [0, 1] \backslash C$ then x belongs to one of the intervals which were deleted in the construction of C; we define $F(x)$ to be the supremum of the set $\{F(y) : y \in C, y < x\}$. Finally, we set $F(x) = 0$ if $x < 0$ and $F(x) = 1$ if $x > 1$. It is fairly easy to see that F is a continuous non-decreasing function from $\mathbb{R}$ onto $[0, 1]$, and thus F is a distribution function.

Let X be a random variable with distribution function F. Clearly X is not a discrete random variable, since F is continuous. It is not quite so easy to see that X cannot be continuous. Roughly speaking, this is because F is constant on each of the intervals $(\frac{1}{3}, \frac{2}{3})$, $(\frac{1}{9}, \frac{2}{9})$, $(\frac{7}{9}, \frac{8}{9})$, ... which were deleted in constructing C; on the other hand the total length of these intervals is

$$\tfrac{1}{3} + 2 \cdot \tfrac{1}{9} + 4 \cdot \tfrac{1}{27} + \cdots = \tfrac{1}{3} \sum_{i=0}^{\infty} (\tfrac{2}{3})^i$$

$$= 1,$$

so that $F'(x) = 0$ for 'almost all' of $[0, 1]$. Thus, if F were to have density function f then $f(x) = 0$ for 'almost all' x, giving that

$$P(-\infty < X < \infty) = \int_{-\infty}^{\infty} f(x) \, dx = 0,$$

which is clearly absurd. Hence, X is neither discrete nor continuous.

It turns out that the distribution function F is in an entirely new category, called the set of 'singular' distribution functions. Do not be too disturbed by this novelty; there are basically only three classes of distribution functions: those which are singular and those which arise from discrete random variables and from continuous random variables.   ☐

Exercises 1. On the kth toss of a fair coin a gambler receives 0 if it is a tail and $2/3^k$ if it is a head. Let X be the total gain of the gambler after an infinite sequence of tosses of the coin. Show that X has the Cantor distribution.
2. Show that the Cantor set is uncountable.

7.2 Moments

The main purpose of this chapter is to study the 'moments' of a random variable: what are moments, and how can we use them? For any random variable X, the kth *moment* of X is defined for $k = 1$, $2, \ldots$ to be the number $E(X^k)$, the expectation of the kth power of X whenever this expectation exists. We shall see that the sequence $E(X)$, $E(X^2)$, $\ldots$ contains a lot of information† about X; but first we give some examples of calculations of moments.

Example 2 *Exponential distribution.* If X has the exponential distribution with parameter λ, then

$$E(X^k) = \int_0^\infty x^k \lambda e^{-\lambda x} \, dx \qquad \text{by} \quad (5.36)$$

$$= \left[-x^k e^{-\lambda x}\right]_0^\infty + \int_0^\infty k x^{k-1} e^{-\lambda x} \, dx$$

$$= \frac{k}{\lambda} E(X^{k-1})$$

if $k \geq 1$, giving that

$$E(X^k) = \frac{k}{\lambda} E(X^{k-1})$$

$$= \frac{k(k-1)}{\lambda^2} E(X^{k-2}) = \cdots$$

$$= \frac{k!}{\lambda^k} E(X^0)$$

$$= \frac{k!}{\lambda^k} E(1) = \frac{k!}{\lambda^k}$$

† Strictly speaking, these moments are associated with the *distribution* of X rather than with X itself; thus we shall speak of the *moments* of a distribution or of a density function.

for $k = 1, 2, \ldots$. Thus, the exponential distribution has moments of all orders. ∎

Example 3

Cauchy distribution. If X has the Cauchy distribution then

$$E(X^k) = \int_{-\infty}^{\infty} \frac{x^k}{\pi(1 + x^2)} \, dx$$

for values of k for which this integral converges absolutely. It is however an elementary exercise (remember Example 5.43) to see that

$$\int_{-\infty}^{\infty} \left| \frac{x^k}{\pi(1 + x^2)} \right| \, dx = \infty$$

if $k \geq 1$, and so the Cauchy distribution possesses *no* moments.

You may see how to adapt this example to find a density function with some moments but not all. Consider the density function

$$f(x) = \frac{c}{1 + |x|^m} \qquad \text{for } x \in \mathbb{R},$$

where m (≥ 2) is an integer, and c is chosen so that f is indeed a density function:

$$c = \left(\int_{-\infty}^{\infty} \frac{dx}{1 + |x|^m} \right)^{-1}.$$

You may check that this density function has a kth moment for those values of k satisfying $1 \leq k \leq m - 2$ only. ∎

Given the distribution function F_X of the random variable, we may calculate the moments of X whenever they exist (at least, if X is discrete or continuous). It is interesting to ask whether or not the converse is true: given the sequence $E(X), E(X^2), \ldots$ of (finite) moments of X, is it possible to reconstruct the distribution of X? The general answer to this question is NO, but is YES if we have some extra information about the moment sequence.

Theorem 7A

Suppose that all moments $E(X), E(X^2), \ldots$ of the random variable X exist, and that the series

(4)
$$\sum_{k=0}^{\infty} \frac{1}{k!} t^k E(X^k)$$

is absolutely convergent for some $t > 0$. Then the sequence of moments uniquely determines the distribution of X.

Thus the absolute convergence of (4) for some $t > 0$ is sufficient (but not necessary) for the moments to determine the underlying distribution. We omit the proof of this, since it is not primarily a theorem about probability theory; a proof may be found in textbooks on real and complex analysis. It is closely related to the uniqueness theorem (Theorem 4A) for probability generating functions; the series in (4) is the exponential generating function of the sequence of moments.

Here is an example of a distribution which is not determined uniquely by its moments.

Example 5

Log-normal distribution. If X has the normal distribution with mean 0 and variance 1, then $Y = e^X$ has the *log-normal distribution* with density function

$$f(x) = \begin{cases} \dfrac{1}{x\sqrt{(2\pi)}} \exp[-\tfrac{1}{2}(\log x)^2] & \text{if } x > 0, \\ 0 & \text{if } x \le 0. \end{cases}$$

Suppose that $-1 \le a \le 1$ and define

$$f_a(x) = [1 + a \sin(2\pi \log x)] f(x).$$

It is not difficult to check that
(a) f_a is a density function,
(b) f has finite moments of all orders,
(c) f_a and f have equal moments of all orders, in that

$$\int_{-\infty}^{\infty} x^k f(x)\, dx = \int_{-\infty}^{\infty} x^k f_a(x)\, dx \qquad \text{for } k = 1, 2, \ldots.$$

Thus $\{f_a : -1 \le a \le 1\}$ is a collection of density functions, each different from all the others but all having the same moments. $\square$

Exercises   3. (i) If X is uniformly distributed on (a, b), show that

$$\mathsf{E}(X^k) = \frac{b^{k+1} - a^{k+1}}{(b-a)(k+1)} \qquad \text{for } k = 1, 2, \ldots.$$

(ii) If X has the gamma distribution with parameters w and λ, show that

$$\mathsf{E}(X^k) = \frac{\Gamma(w+k)}{\lambda^k \Gamma(w)} \qquad \text{for } k = 1, 2, \ldots.$$

(iii) If X has the χ^2 distribution with n degrees of freedom, show that

$$\mathsf{E}(X^k) = 2^k \frac{\Gamma(k + \tfrac{1}{2}n)}{\Gamma(\tfrac{1}{2}n)} \qquad \text{for } k = 1, 2, \ldots.$$

7.3 Variance and covariance

We recall that the *variance* of a random variable X is defined to be

(6) $$\mathrm{var}(X) = \mathsf{E}((X - \mu)^2)$$

where $\mu = E(X)$ (see (2.21) and (5.38) for discrete and continuous random variables). The variance of X is a measure of its dispersion about the expectation μ, in the sense that if X often takes values which differ considerably from μ then $|X - \mu|$ is often large and so $E((X - \mu)^2)$ will be large, whereas if X is usually near to μ then $|X - \mu|$ is usually small and $E((X - \mu)^2)$ is small also. An extreme case arises when X is *concentrated* at some point. It is the case that, for a random variable Y,

(7) $$E(Y^2) = 0 \quad \text{if and only if} \quad P(Y = 0) = 1;$$

obviously $E(Y^2) = 0$ if $P(Y = 0) = 1$, and the converse holds since (for discrete random variables, anyway)

$$E(Y^2) = \sum_y y^2 P(Y = y) \geq 0$$

with equality if and only if $P(Y = y) = 0$ for all $y \neq 0$. Applying (7) to $Y = X - \mu$ gives

(8) $$\text{var}(X) = 0 \quad \text{if and only if} \quad P(X = \mu) = 1,$$

so that 'zero variance' means 'no dispersion at all'.

There are many other possible measures of dispersion, such as $E(|X - \mu|)$ and $E(|X - \mu|^3)$ and so on, but it is easiest to work with variances.

As noted before, when calculating the variance of X it is often easier to work with the moments of X rather than with (6) directly. That is to say, it may be easier to make use of the formula

(9) $$\begin{aligned} \text{var}(X) &= E((X - \mu)^2) \\ &= E(X^2) - \mu^2 \end{aligned}$$

by (2.23) and (5.39), where $\mu = E(X)$.

There is also a simple formula for calculating the variance of a linear function $aX + b$ of a random variable X, namely

(10) $$\text{var}(aX + b) = a^2 \, \text{var}(X).$$

To see this, note that

$$\begin{aligned} \text{var}(aX + b) &= E([aX + b - E(aX + b)]^2) \\ &= E([aX + b - aE(X) - b]^2) \qquad \text{by (6.41)} \\ &= E(a^2(X - \mu)^2) \\ &= a^2 E((X - \mu)^2) \qquad \text{by (6.41)} \\ &= a^2 \, \text{var}(X). \end{aligned}$$

As a measure of dispersion, the variance of X has an undesirable property: it is non-linear in the sense that the variance of aX is a^2 times the variance of X, for $a \in \mathbb{R}$. For this reason, statisticians often prefer to work with the *standard deviation* of X, defined to be $\sqrt{\mathrm{var}(X)}$.

What can be said about $\mathrm{var}(X + Y)$ in terms of $\mathrm{var}(X)$ and $\mathrm{var}(Y)$? It is simple to see that

$$
\begin{aligned}
(11) \quad \mathrm{var}(X + Y) &= \mathsf{E}\big([(X + Y) - \mathsf{E}(X + Y)]^2\big) \\
&= \mathsf{E}\big(\{[X - \mathsf{E}(X)] + [Y - \mathsf{E}(Y)]\}^2\big) \\
&= \mathsf{E}\big([X - \mathsf{E}(X)]^2 + 2[X - \mathsf{E}(X)][Y - \mathsf{E}(Y)] + [Y - \mathsf{E}(Y)]^2\big) \\
&= \mathrm{var}(X) + 2\mathsf{E}\big((X - \mathsf{E}(X))(Y - \mathsf{E}(Y))\big) + \mathrm{var}(Y).
\end{aligned}
$$

It is convenient to have a special word for the middle term in the last expression, and to this end we define the *covariance* $\mathrm{cov}(X, Y)$ of X and Y by

$$(12) \qquad \mathrm{cov}(X, Y) = \mathsf{E}\big([X - \mathsf{E}(X)][Y - \mathsf{E}(Y)]\big),$$

whenever these expectations exist. Note that $\mathrm{cov}(X, Y)$ may be written in a simpler form: expand $[X - \mathsf{E}(X)][Y - \mathsf{E}(Y)]$ in (12) and use the linearity of E to find that

$$(13) \qquad \mathrm{cov}(X, Y) = \mathsf{E}(XY) - \mathsf{E}(X)\mathsf{E}(Y).$$

Note that (11) may be rewritten as

$$(14) \qquad \mathrm{var}(X + Y) = \mathrm{var}(X) + 2\,\mathrm{cov}(X, Y) + \mathrm{var}(Y),$$

valid for all random variables X and Y. If X and Y are independent then

$$(15) \qquad \mathrm{cov}(X, Y) = \mathsf{E}(XY) - \mathsf{E}(X)\mathsf{E}(Y) = 0$$

by (6.42), giving that the sum of independent random variables has variance

$$(16) \qquad \mathrm{var}(X + Y) = \mathrm{var}(X) + \mathrm{var}(Y)$$

whenever the latter variances exist. The converse of the last remark is false in general: we recall from (3.13) that there exist dependent random variables X and Y for which $\mathrm{cov}(X, Y) = 0$. Despite this, $\mathrm{cov}(X, Y)$ is often used as a measure of the dependence of X and Y, and the reason for this is that $\mathrm{cov}(X, Y)$ is a single number (rather than a complicated object such as a joint density function) which contains some useful information about the *joint* behaviour of X and Y: for example, if $\mathrm{cov}(X, Y) > 0$ then $X - \mathsf{E}(X)$ and $Y - \mathsf{E}(Y)$ may have a good chance (in some sense) of having the same sign. A principal disadvantage of covariance as a measure of dependence is that it is not 'scale-invariant': if X and Y are random measurements (in inches say) and U and V are the same random measurements in centimetres (so that $U = \alpha X$ and $V = \alpha Y$ where $\alpha \approx 2.54$) then $\text{cov}(U, V) \approx 6 \, \text{cov}(X, Y)$ despite the fact that the two pairs, (X, Y) and (U, V), measure essentially the same quantities. To deal with this, we 're-scale' covariance by defining the *correlation (coefficient)* of random variables X and Y to be the quantity $\rho(X, Y)$ given by

(17)
$$\rho(X, Y) = \frac{\text{cov}(X, Y)}{\sqrt{[\text{var}(X) \, \text{var}(Y)]}},$$

whenever the latter quantities exist and $\text{var}(X) \, \text{var}(Y) \neq 0$.

It is a simple exercise to show that

(18)
$$\rho(aX + b, cY + d) = \rho(X, Y)$$

for all $a, b, c, d \in \mathbb{R}$ such that $ac \neq 0$, and so correlation is scale-invariant. Correlation has another attractive property as a measure of dependence. It turns out that $-1 \leq \rho(X, Y) \leq 1$ always, and moreover there are specific interpretations in terms of the joint behaviour of X and Y of the cases when $\rho(X, Y) = \pm 1$.

Theorem 7B *If X and Y are random variables, then*

(19)
$$-1 \leq \rho(X, Y) \leq 1,$$

whenever this correlation exists.

The proof of this is a direct application of the next inequality.

Theorem 7C *Cauchy–Schwarz inequality. If U and V are random variables, then*

(20)
$$[E(UV)]^2 \leq E(U^2) E(V^2),$$

whenever these expectations exist.

Proof Let $s \in \mathbb{R}$ and define a new random variable $W = sU + V$. Clearly $W^2 \geq 0$ always, and so

(21)
$$0 \leq E(W^2)$$
$$= E(s^2 U^2 + 2sUV + V^2)$$
$$= as^2 + bs + c$$

where $a = E(U^2)$, $b = 2E(UV)$, $c = E(V^2)$. Clearly $a \geq 0$, and we may suppose that $a > 0$, since otherwise $P(U = 0) = 1$ by (7) and the result holds trivially. Equation (21) implies that the quadratic function $g(s) = as^2 + bs + c$ intersects the line $t = 0$ (in the usual (s, t) plane) at most once (since if $g(s) = 0$ for distinct values $s = s_1$ and $s = s_2$ of s, then $g(s) < 0$ for all values of s strictly between s_1 and s_2). Thus, the quadratic equation '$g(s) = 0$' has at most one real root, giving that its discriminant $b^2 - 4ac$ satisfies $b^2 - 4ac \leq 0$. Hence

$$[2E(UV)]^2 - 4E(U^2)E(V^2) \leq 0$$

and the result is proved. □

Proof of Theorem 7B Set $U = X - E(X)$ and $V = Y - E(Y)$ in the Cauchy–Schwarz inequality to find that

$$[cov(X, Y)]^2 \leq var(X) \, var(Y),$$

yielding (19) immediately. □

Only under very special circumstances can it be the case that $\rho(X, Y) = \pm 1$, and these circumstances are explored by considering the proof of (20) more carefully. Suppose that $a = var(X)$, $b = 2 \, cov(X, Y)$, $c = var(Y)$ and that $\rho(X, Y) = \pm 1$. Then $var(X) \, var(Y) \neq 0$ and

$$b^2 - 4ac = 4 \, var(X) \, var(Y)[\rho(X, Y)^2 - 1] = 0,$$

and so the quadratic equation

$$as^2 + bs + c = 0$$

has two equal real roots, at $s = \alpha$ say. Hence the random variable $W = \alpha(X - E(X)) + (Y - E(Y))$ satisfies

$$E(W^2) = a\alpha^2 + b\alpha + c = 0,$$

giving that $P(W = 0) = 1$ by (7), and showing that (essentially) $Y = -\alpha X + \beta$ where $\beta = \alpha E(X) + E(Y)$. A slightly more careful treatment discriminates between the values $+1$ and -1 for $\rho(X, Y)$:

(22) $$\rho(X, Y) = 1 \text{ if and only if } P(Y = \alpha X + \beta) = 1$$

for some real α and β with $\alpha > 0$,

(23) $$\rho(X, Y) = -1 \text{ if and only if } P(Y = \alpha X + \beta) = 1$$

for some real α and β with $\alpha < 0$.

To recap, we may use $\rho(X, Y)$ as a measure of the dependence of X and Y. If X and Y have non-zero variances, then $\rho(X, Y)$ takes some value in the interval $[-1, 1]$, and this value should be interpreted in the light of the ways in which the values -1, 0, 1 may arise:

(a) if X and Y are independent then $\rho(X, Y) = 0$,
(b) Y is a linear increasing function of X if and only if $\rho(X, Y) = 1$,
(c) Y is a linear decreasing function of X if and only if $\rho(X, Y) = -1$.

If $\rho(X, Y) = 0$, we say that X and Y are *uncorrelated*.

Exercises   4. If X and Y have the bivariate normal distribution with parameters μ_1, μ_2, σ_1, σ_2, ρ (see (6.51)), show that

$$\text{cov}(X, Y) = \rho\sigma_1\sigma_2 \quad \text{and} \quad \rho(X, Y) = \rho.$$

5. Let $X_1, X_2, \ldots$ be a sequence of uncorrelated random variables each having variance σ^2. If $S_n = X_1 + X_2 + \cdots + X_n$, show that

$$\text{cov}(S_n, S_m) = \text{var}(S_n) = n\sigma^2 \quad \text{if } n < m.$$

6. Show that $\text{cov}(X, Y) = 1$ in the case when X and Y have joint density function

$$f(x, y) = \begin{cases} \dfrac{1}{y} e^{-y-x/y} & \text{if } x, y > 0, \\ 0 & \text{otherwise.} \end{cases}$$

7.4 Moment generating functions

If X is a discrete random variable taking values in $\{0, 1, 2, \ldots\}$, its probability generating function is defined by

(24)
$$G_X(s) = \mathsf{E}(s^X) = \sum_{k=0}^{\infty} s^k \mathsf{P}(X = k).$$

Probability generating functions are very useful, but only when the random variables in question take non-negative integral values. For more general random variables, it is customary to consider a modification of (24). The *moment generating function* of a random variable X is the function M_X defined by

(25)
$$M_X(t) = \mathsf{E}(e^{tX})$$

for all $t \in \mathbb{R}$ for which this expectation exists. This is a modification of (24) in the sense that if X takes values in $\{0, 1, 2, \ldots\}$ then

(26)
$$M_X(t) = \mathsf{E}(e^{tX}) = G_X(e^t),$$

by substituting $s = e^t$ in (24). In general

(27)
$$M_X(t) = \mathsf{E}(e^{tX}) = \begin{cases} \displaystyle\sum_x e^{tx} \mathsf{P}(X = x) & \text{if } X \text{ is discrete,} \\ \displaystyle\int_{-\infty}^{\infty} e^{tx} f_X(x)\, dx & \text{if } X \text{ is continuous with density function } f_X, \end{cases}$$

whenever this sum or integral converges absolutely. In some cases the existence of $M_X(t)$ can be a problem for non-zero values of t.

Example 28

Normal distribution. If X has the normal distribution with mean 0 and variance 1 then

(29)
$$M_X(t) = \int_{-\infty}^{\infty} e^{tx} \frac{1}{\sqrt{(2\pi)}} e^{-\frac{1}{2}x^2} \, dx$$

$$= e^{\frac{1}{2}t^2} \int_{-\infty}^{\infty} \frac{1}{\sqrt{(2\pi)}} e^{-\frac{1}{2}(x-t)^2} \, dx$$

$$= e^{\frac{1}{2}t^2},$$

since the integrand in the latter integral is the density function of the normal distribution with mean t and variance 1, and thus has integral 1. The moment generating function $M_X(t)$ exists for all $t \in \mathbb{R}$. □

Example 30

Exponential distribution. If X has the exponential distribution with parameter λ, then

(31)
$$M_X(t) = \int_0^{\infty} e^{tx} \lambda e^{-\lambda x} \, dx$$

$$= \begin{cases} \dfrac{\lambda}{\lambda - t} & \text{if } t < \lambda, \\ \infty & \text{if } t \geq \lambda, \end{cases}$$

so that $M_X(t)$ exists only for values of t satisfying $t < \lambda$. □

Example 32

Cauchy distribution. If X has the Cauchy distribution then

$$M_X(t) = \int_{-\infty}^{\infty} e^{tx} \frac{1}{\pi(1 + x^2)} \, dx$$

$$= \begin{cases} 1 & \text{if } t = 0, \\ \infty & \text{if } t \neq 0, \end{cases}$$

so that $M_X(t)$ exists only at $t = 0$. □

This difficulty over the existence of $E(e^{tX})$ may be avoided by studying the complex-valued *characteristic function* $\phi_X(t) = E(e^{itX})$ of X instead—this function can be shown to exist for all $t \in \mathbb{R}$. However, we want to avoid $i = \sqrt{(-1)}$ at this stage, and so we must accustom ourselves to the difficulty although we shall return to characteristic functions in Section 7.5. It turns out to be important only that $E(e^{tX})$ exists in some neighbourhood $(-\delta, \delta)$ of the origin, and the reason for this is contained in the uniqueness theorem for moment generating functions; we defer this until the end of the section. We shall generally use moment generating functions freely, but always subject to the implicit assumption of existence in a neighbourhood of the origin.

The reason for the name 'moment generating function' is the following intuitively attractive expansion:

(33)
$$M_X(t) = E(e^{tX})$$

$$= E\left(1 + tX + \frac{1}{2!}(tX)^2 + \cdots\right)$$

$$= 1 + tE(X) + \frac{1}{2!}t^2E(X^2) + \cdots.$$

That is to say, subject to a rigorous derivation of (33) which does not interchange the two operations E and Σ so light-heartedly, $M_X(t)$ is the exponential generating function of the moments of X.

Theorem 7D   *If $M_X(t)$ exists in a neighbourhood of 0 then, for $k = 1, 2, \ldots$,*

(34)
$$E(X^k) = M_X^{(k)}(0),$$

the kth derivative of $M_X(t)$ evaluated at $t = 0$.

Sketch proof   Cross your fingers for the sake of rigour to obtain

$$\frac{d^k}{dt^k} M_X(t) = \frac{d^k}{dt^k} E(e^{tX})$$

$$= E\left(\frac{d^k}{dt^k} e^{tX}\right)$$

$$= E(X^k e^{tX}),$$

and finish by setting $t = 0$. It is the interchange of the expectation operator and the differential operator which requires justification here.   □

As noted before, much of probability theory is concerned with sums of random variables. It can be difficult in practice to calculate the distribution of a sum from knowledge of the distributions of the summands, and it is here that moment generating functions are supremely useful. Consider first the linear function $aX + b$ of the random variable X. If $a, b \in \mathbb{R}$ then

$$M_{aX+b}(t) = E(e^{t(aX+b)})$$

$$= E(e^{atX}e^{tb})$$

$$= e^{tb}E(e^{(at)X}) \qquad \text{by (6.41)}$$

giving that

(35)
$$M_{aX+b}(t) = e^{tb}M_X(at).$$

A similar argument enables us to find the moment generating function of the sum of independent random variables.

Theorem 7E *If X and Y are independent random variables then $X + Y$ has moment generating function*

(36)
$$M_{X+Y}(t) = M_X(t)M_Y(t).$$

Proof
$$M_{X+Y}(t) = \mathsf{E}(e^{t(X+Y)})$$
$$= \mathsf{E}(e^{tX}e^{tY})$$
$$= \mathsf{E}(e^{tX})\mathsf{E}(e^{tY})$$

by independence and Theorem 6F. □

It follows from (36) that the sum $S = X_1 + \cdots + X_n$ of n independent random variables has moment generating function

(37)
$$M_S(t) = M_{X_1}(t) \cdots M_{X_n}(t).$$

Finally, we state the uniqueness theorem for moment generating functions.

Theorem 7F *If $M_X(t) = \mathsf{E}(e^{tX}) < \infty$ for all t satisfying $-\delta < t < \delta$ and some $\delta > 0$, then there is a unique distribution with moment generating function M_X. Furthermore, under this condition, we have that $\mathsf{E}(X^k) < \infty$ for $k = 1, 2, \ldots$ and*

(38)
$$M_X(t) = \sum_{k=0}^{\infty} \frac{1}{k!} t^k \mathsf{E}(X^k) \qquad for \ |t| < \delta.$$

We do not prove this here. This theorem is basically the Laplace inverse theorem since, by (27), $M_X(t)$ is essentially the Laplace transform of the density function $f_X(x)$; the Laplace inverse theorem says that if the Laplace transform of f_X exists in a suitable manner then f_X may be found from this transform by using the inversion formula. Equation (38) is the same as (33), but some care is needed to justify the interchange of E and Σ noted after (33). Clearly there is a close relationship between Theorem 7F and Theorem 7A, but we do not explore this here.

Finally we give an example of the use of moment generating functions in which the uniqueness part of Theorem 7F is essential.

Example 39 Let X and Y be independent random variables, X having the normal distribution with parameters μ_1 and σ_1^2 and Y having the normal distribution with parameters μ_2 and σ_2^2. Show that their sum $Z = X + Y$ has the normal distribution with parameters $\mu_1 + \mu_2$ and $\sigma_1^2 + \sigma_2^2$.

Solution   Let U be a random variable having the normal distribution with parameters μ and σ^2. The moment generating function of U is

(40)
$$M_U(t) = \int_{-\infty}^{\infty} e^{tu} \frac{1}{\sqrt{(2\pi\sigma^2)}} \exp\left(-\frac{1}{2\sigma^2}(u-\mu)^2\right) du$$

$$= e^{\mu t} \int_{-\infty}^{\infty} e^{x\sigma t} \frac{1}{\sqrt{(2\pi)}} e^{-\frac{1}{2}x^2} dx \qquad \text{by the substitution } x = \frac{u-\mu}{\sigma}$$

$$= \exp(\mu t + \tfrac{1}{2}\sigma^2 t^2) \qquad \text{by (29)}.$$

By Theorem 7E,

$$M_Z(t) = M_X(t)M_Y(t)$$
$$= \exp(\mu_1 t + \tfrac{1}{2}\sigma_1^2 t^2)\exp(\mu_2 t + \tfrac{1}{2}\sigma_2^2 t^2) \qquad \text{by (40)}$$
$$= \exp[(\mu_1 + \mu_2)t + \tfrac{1}{2}(\sigma_1^2 + \sigma_2^2)t^2],$$

and we recognize this by (40) as being the moment generating function of the normal distribution with parameters $\mu_1 + \mu_2$ and $\sigma_1^2 + \sigma_2^2$; we deduce that Z has this distribution by appealing to Theorem 7F.   □

Exercises   7. Find the moment generating function of a random variable having
  (i) the gamma distribution with parameters w and λ,
  (ii) the Poisson distribution with parameter λ.
  8. If X has the normal distribution with mean μ and variance σ^2, find $E(X^3)$.
  9. Use Theorem 7F together with (35) and (40) to show that if X has a normal distribution then so does $aX + b$, for any $a, b \in \mathbb{R}$ with $a \neq 0$.

7.5   Characteristic functions†

The Cauchy distribution is not the only distribution for which the moment generating function does not exist, and this problem of existence is a serious handicap to the use of moment generating functions. However, by a slight modification of the definition, we may obtain another type of generating function whose existence is guaranteed and which has broadly the same properties as before.

The *characteristic function* of a random variable X is defined to be the function ϕ_X given by

(41)
$$\phi_X(t) = E(e^{itX}) \qquad \text{for } t \in \mathbb{R},$$

where $i = \sqrt{(-1)}$. You may doubt the legitimacy of the expectation of the complex-valued random variable e^{itX}, but we recall that $e^{itx} =$

† Beginners to probability theory may wish to omit this section.

$\cos tx + i \sin tx$ for $t, x \in \mathbb{R}$, so that (41) may be replaced by

$$\phi_X(t) = E(\cos tX) + iE(\sin tX)$$

if this is preferred.

Compare the characteristic function of X with its moment generating function $M_X(t) = E(e^{tX})$. The finiteness of the latter is questionable since the exponential function is unbounded, so that e^{tX} may be very large indeed. On the other hand e^{itX} lies on the unit circle in the complex plane, so that $|e^{itX}| = 1$ and giving that $|\phi_X(t)| \leq 1$ for all $t \in \mathbb{R}$.

Example 42

Suppose that the random variable X may take either the value a, with probability p, or the value b, with probability $1 - p$. Then

$$\phi_X(t) = E(e^{itX})$$
$$= pe^{ita} + (1-p)e^{itb}. \qquad \square$$

Example 43

Exponential distribution. If X has the exponential distribution with parameter λ, then

$$\phi_X(t) = \int_0^\infty e^{itx} \lambda e^{-\lambda x} \, dx$$

$$= \frac{\lambda}{\lambda - it} \qquad \text{for } t \in \mathbb{R}.$$

This integral may be found either by splitting e^{itx} into real and imaginary parts or by using the calculus of residues. $\qquad \square$

Example 44

Cauchy distribution. If X has the Cauchy distribution, then

$$\phi_X(t) = \int_{-\infty}^\infty e^{itx} \frac{1}{\pi(1+x^2)} \, dx$$

$$= e^{-|t|} \qquad \text{for } t \in \mathbb{R},$$

a result obtainable by the calculus of residues. $\qquad \square$

Some readers may prefer to avoid using the calculus of residues in calculating characteristic functions, arguing instead as in the following example. The moment generating function of a random variable X having the normal distribution with mean 0 and variance 1 is

$$M_X(t) = E(e^{tX}) = e^{\frac{1}{2}t^2}.$$

'It is therefore clear that the characteristic function of X is

$$\phi_X(t) = E(e^{itX})$$
$$= M_X(it) = e^{-\frac{1}{2}t^2} \qquad \text{for } t \in \mathbb{R}.'$$

It is important to realize that this argument is not rigorous unless justified. It produces the correct answer for the normal and exponential distributions, as well as many others, but it will not succeed with the Cauchy distribution, since that distribution has no moment generating function in the first place. This argument may be shown to be valid whenever the moment generating function exists near the origin, the proof being an exercise in complex analysis. Thus the following formal procedure is acceptable for calculating the characteristic function of a random variable X. If the moment generating function M_X of X is finite in a non-trivial neighbourhood of the origin, the characteristic function of X may be found by substituting $s = it$ in the formula for $M_X(s)$:

(45) $$\phi_X(t) = M_X(it) \qquad \text{for } t \in \mathbb{R}.$$

Example 46   *Normal distribution.* If X has the normal distribution with mean μ and variance σ^2, then the moment generating function of X is

$$M_X(s) = \exp(\mu s + \tfrac{1}{2}\sigma^2 s^2)$$

by (40), an expression valid for all $s \in \mathbb{R}$. We substitute $s = it$ here, to obtain

$$\phi_X(t) = \exp(i\mu t - \tfrac{1}{2}\sigma^2 t^2). \qquad \square$$

In broad terms, characteristic functions have the same useful properties as moment generating functions, and we finish this chapter with a brief account of these.

First we consider the question of moments. Setting rigour to one side for the moment, the following expansion is interesting and informative:

(47) $$\phi_X(t) = E(e^{itX})$$

$$= E\left(1 + itX + \frac{1}{2!}(itX)^2 + \cdots\right)$$

$$= 1 + it E(X) + \frac{1}{2!}(it)^2 E(X^2) + \cdots,$$

which is to say that ϕ_X is the exponential generating function of the sequence $1, iE(X), i^2 E(X^2), \ldots$. There are technical difficulties in expressing this more rigorously, but we note that (47) is valid so long as $E|X^k| < \infty$ for $k = 1, 2, \ldots$. Under this condition, it follows that the moments of X may be obtained in terms of the derivatives of ϕ_X:

(48) $$i^k E(X^k) = \phi_X^{(k)}(0),$$

the kth derivative of ϕ_X at 0.

If the moments of X are not all finite, then only a truncated form of the infinite series in (47) is valid.

Theorem 7G

If $E|X^N| < \infty$ *for some positive integer* N, *then*

(49)
$$\phi_X(t) = \sum_{k=0}^{N} \frac{1}{k!} (it)^k E(X^k) + o(t^N).$$

We do not prove this here, but remind the reader briefly about the meaning of the term $o(t^N)$. The expression $o(h)$ denotes some function of h which is of smaller order of magnitude than h as $h \to 0$. More precisely, we write $f(h) = o(h)$ if $h^{-1}f(h) \to 0$ as $h \to 0$. The term $o(h)$ generally represents a different function of h at each appearance. Thus, for example, $o(h) + o(h) = o(h)$. The conclusion of this theorem is that the remainder in (49) is negligible compared with the terms involving 1, t, t^2, ..., t^N. For a proof of Theorem 7G, see Feller (1971, p. 487) or Chung (1974, p. 168).

When adding together independent random variables, characteristic functions are just as useful as moment generating functions.

Theorem 7H

Let X and Y be independent random variables with characteristic functions ϕ_X and ϕ_Y respectively.
(a) If a, $b \in \mathbb{R}$ and $Z = aX + b$ then

$$\phi_Z(t) = e^{itb} \phi_X(at).$$

(b) The characteristic function of $X + Y$ is

$$\phi_{X+Y}(t) = \phi_X(t)\phi_Y(t).$$

Proof   (a)
$$\phi_Z(t) = E(e^{it(aX+b)})$$
$$= E(e^{itb} e^{i(at)X})$$
$$= e^{itb} \phi_X(at).$$

If you are in doubt about treating these complex-valued quantities as if they were real, simply expand the exponential function in terms of the cosine and sine functions, and collect the terms together again at the end.
(b) Similarly,

$$\phi_{X+Y}(t) = E(e^{it(X+Y)})$$
$$= E(e^{itX} e^{itY})$$
$$= E(e^{itX})E(e^{itY}) \qquad \text{by independence.} \qquad \square$$

Finally, we discuss the uniqueness of characteristic functions. The basic uniqueness theorem is of great value.

Theorem 7I   *Let X and Y have characteristic functions ϕ_X and ϕ_Y respectively. Then X and Y have the same distributions if and only if $\phi_X(t) = \phi_Y(t)$ for all $t \in \mathbb{R}$.*

That is to say, any given characteristic function ϕ corresponds to a unique distribution function. However, it is not always a simple matter to find this distribution function in terms of ϕ. There is a general 'inversion formula', but this is rather complicated and so we omit this here (see Grimmett and Stirzaker (1982, p. 106)). For distributions with density functions, the inversion formula takes on a relatively simple form.

Theorem 7J   *Let X have characteristic function ϕ and density function f. Then*

(50)
$$f(x) = \frac{1}{2\pi} \int_{-\infty}^{\infty} e^{-itx} \phi(t) \, dt$$

at every point x at which f is differentiable.

This formula is often useful, but there is an obstacle in the way of its application. If we are given a characteristic function ϕ, we may only apply formula (50) once we know that ϕ comes from a continuous random variable; but how may we check that this is the case? There is no attractive necessary and sufficient condition on ϕ for this to hold, but a sufficient condition is that

(51)
$$\int_{-\infty}^{\infty} |\phi(t)| \, dt < \infty.$$

This condition is only of limited value: although it holds for the characteristic function of the normal distribution (46) for example, it fails for that of the exponential distribution (43).

Example 52   Those in the know will have spotted that characteristic functions are simply Fourier transforms in disguise, and that Theorem 7J is a version of the Fourier inverse theorem. The relationship between characteristic functions and Fourier analysis may easily be made more concrete in the case of integer-valued random variables. Suppose that X is a random variable taking values in the non-negative integers $\{0, 1, 2, \ldots\}$ with probability mass function $p(k) = P(X = k)$ for $k = 0, 1, 2, \ldots$. The characteristic function of X is

(53)
$$\phi(t) = \sum_{k=0}^{\infty} p(k) e^{itk}.$$

Suppose now that we know ϕ, but wish to recover the probabilities

$p(k)$. We multiply through (53) by e^{-itj} to obtain

$$e^{-itj}\phi(t) = \sum_{k=0}^{\infty} p(k)e^{it(k-j)}.$$

Next we integrate over t from 0 to 2π, remembering that for integers m

$$\int_0^{2\pi} e^{imt}\,dt = \begin{cases} 2\pi & \text{if } m = 0, \\ 0 & \text{if } m \neq 0, \end{cases}$$

and obtain

$$\int_0^{2\pi} e^{-itj}\phi(t)\,dt = 2\pi p(j),$$

so that

(54)
$$p(j) = \frac{1}{2\pi}\int_0^{2\pi} e^{-itj}\phi(t)\,dt \qquad \text{for } j = 0, 1, 2, \ldots.$$

We are merely calculating the Fourier series for ϕ. Notice the close resemblance between (54) and the inversion formula (50) for density functions. □

Exercises
10. Show that the characteristic function of a random variable having the binomial distribution with parameters n and p is

$$\phi(t) = (q + pe^{it})^n,$$

where $q = 1 - p$.
11. Let X be uniformly distributed on (a, b). Show that

$$\phi_X(t) = \frac{e^{itb} - e^{ita}}{it(b - a)}.$$

If X is uniformly distributed on $(-b, b)$, show that

$$\phi_X(t) = \frac{1}{bt}\sin bt.$$

12. Find the characteristic function of a random variable having the
    (i) gamma distribution with parameters w and λ,
    (ii) the Poisson distribution with parameter λ.
13. If X and Y are independent and identically distributed random variables, show that

$$\phi_{X-Y}(t) = |\phi_X(t)|^2.$$

7.6 Problems

1. Let X and Y be random variables with equal variance. Show that $U = X - Y$ and $V = X + Y$ are uncorrelated. Give an example to show that U and V need not be independent even if X and Y are independent.

2. Let $X_1, X_2, \ldots$ be uncorrelated random variables each having mean μ and variance σ^2. If $\bar{X} = n^{-1}(X_1 + X_2 + \cdots + X_n)$, show that

$$E\left(\frac{1}{n-1}\sum_{i=1}^{n}(X_i - \bar{X})^2\right) = \sigma^2.$$

(This fact is of great importance in statistics and is used when estimating the population variance from knowledge of a random sample.)

3. Let $X_1, X_2, \ldots$ be identically distributed independent random variables and let $S_n = X_1 + X_2 + \cdots + X_n$. Show that

$$E\left(\frac{S_m}{S_n}\right) = \frac{m}{n} \qquad \text{for } m \leq n,$$

provided that all the necessary expectations exist. Is the same true if $m > n$?

4. Show that every distribution function has only a countable set of points of discontinuity.

5. Let X and Y be independent random variables, X having the gamma distribution with parameters s and λ, and Y having the gamma distribution with parameters t and λ. Use moment generating functions to show that $X + Y$ has the gamma distribution with parameters $s + t$ and λ.

6. Let $X_1, X_2, \ldots, X_n$ be independent random variables with the exponential distribution, parameter λ. Show that $X_1 + X_2 + \cdots + X_n$ has the gamma distribution with parameters n and λ.

7. Show from the result of Problem 5 that the χ^2 distribution with n degrees of freedom has moment generating function

$$M(t) = (1 - 2t)^{-\frac{1}{2}n} \qquad \text{if } t < \tfrac{1}{2}.$$

Deduce that, if $X_1, X_2, \ldots, X_n$ are independent random variables having the normal distribution with mean 0 and variance 1, then

$$Z = X_1^2 + X_2^2 + \cdots + X_n^2$$

has the χ^2 distribution with n degrees of freedom. Hence or otherwise show that the sum of two independent random variables, having the χ^2 distribution with m and n degrees of freedom respectively, has the χ^2 distribution with $m + n$ degrees of freedom.

8. Let $X_1, X_2, \ldots$ be independent, identically distributed random variables and let N be a random variable which takes values in the positive integers and is independent of the X's. Find the moment generating function of

$$S = X_1 + X_2 + \cdots + X_N$$

in terms of the moment generating functions of N and X_1, when these exist.

9. Random variables $X_1, X_2, \ldots, X_N$ have zero expectations, and

$$E(X_m X_n) = v_{mn} \qquad (m, n = 1, 2, \ldots, N).$$

Calculate the variance of the random variable

$$Z = \sum_{n=1}^{N} a_n X_n,$$

and deduce that the symmetric matrix $V = [v_{mn}]$ is non-negative-definite. It is desired to find an $N \times N$ matrix A such that the random variables

$$Y_n = \sum_{r=1}^{N} a_{nr}X_r \quad (n = 1, 2, \ldots, N)$$

are uncorrelated, and have unit variance. Show that this will be the case if and only if

$$AVA^T = I,$$

and show that A can be chosen to satisfy this equation if and only if V is non-singular.

(Any standard results from matrix theory may, if clearly stated, be used without proof. A^T denotes the transpose of A.) (Oxford 1971F)

10. Prove that if $X = X_1 + \cdots + X_n$ and $Y = Y_1 + \cdots + Y_n$ where X_i and Y_j are independent whenever $i \neq j$, then $\text{cov}(X, Y) = \sum_{i=1}^{n} \text{cov}(X_i, Y_i)$. (Assume that all series involved are absolutely convergent.)

Two players A and B play a series of independent games. The probability that A wins any particular game is p^2, that B wins is q^2, and that the game is a draw is $2pq$, where $p + q = 1$. The winner of a game scores 2 points, the loser none; if a game is drawn, each player scores 1 point. Let X and Y be the number of points scored by A and B, respectively, in a series of n games. Prove that $\text{cov}(X, Y) = -2npq$.
(Oxford 1982M)

11. The *joint moment generating function* of two random variables X and Y is defined to be the function $M(s, t)$ of two real variables defined by

$$M(s, t) = E(e^{sX + tY})$$

for all values of s and t for which this expectation exists. Show that the joint moment generating function of a pair of random variables having the standard bivariate normal distribution (6.48) is

$$M(s, t) = \exp[\tfrac{1}{2}(s^2 + 2\rho st + t^2)].$$

Deduce the joint moment generating function of a pair of random variables having the bivariate normal distribution (6.51) with parameters μ_1, μ_2, σ_1, σ_2, ρ.

*12. Let X and Y be independent random variables, each having mean 0, variance 1, and finite moment generating function $M(t)$. If $X + Y$ and $X - Y$ are independent, show that

$$M(2t) = M(t)^3 M(-t)$$

and deduce that X and Y have the normal distribution with mean 0 and variance 1.

13. Let X have moment generating function $M(t)$.
 (i) Show that $M(t)M(-t)$ is the moment generating function of $X - Y$ where Y is independent of X but has the same distribution.

In the same way, describe random variables which have moment generating function

 (ii) $[2 - M(t)]^{-1}$,   (iii) $\int_0^\infty M(ut)e^{-u} \, du$.

14. *Coupon-collecting again.* There are c different types of coupon and each time you obtain a coupon it is equally likely to be any one of the c types. Find the moment generating function of the total number N of coupons which you collect in order to obtain a complete set.

15. Prove that if ϕ_1 and ϕ_2 are characteristic functions then so is $\phi = \alpha\phi_1 + (1 - \alpha)\phi_2$ for any $\alpha \in \mathbb{R}$ satisfying $0 \le \alpha \le 1$.

16. Show that X and $-X$ have the same distribution if and only if ϕ_X is a purely real-valued function.

17. Find the characteristic function of a random variable with density function

$$f(x) = \tfrac{1}{2}e^{-|x|} \quad \text{for } x \in \mathbb{R}.$$

18. Let $X_1, X_2, \ldots$ be independent random variables each having the Cauchy distribution, and let

$$S_n = \frac{1}{n}(X_1 + X_2 + \cdots + X_n).$$

Show that S_n has the Cauchy distribution regardless of the value of n.

19. Show that $\phi(t) = \exp(-|t|^\alpha)$ is the characteristic function of a distribution with finite variance if and only if $\alpha = 2$.

20. Let X be a random variable whose moment generating function $M(t)$ exists for $|t| < h$. Let N be a random variable taking positive integer values such that

$$P(N = k) > 0 \quad \text{for } k = 1, 2, \ldots.$$

Show that

$$M(t) = \sum_{k=1}^{\infty} P(N = k)E(e^{tX} \mid N = k) \quad \text{for } |t| < h.$$

Let $X = \max(U_1, U_2, \ldots, U_N)$ where the U_i are independent random variables uniformly distributed on $(0, 1)$ and N is an independent random variable whose distribution is given by

$$P(N = k) = [(e - 1)k!]^{-1} \quad \text{for } k = 1, 2, \ldots.$$

Obtain the moment generating function of X and hence show that if R is another independent random variable with

$$P(R = r) = (e - 1)e^{-r} \quad \text{for } r = 1, 2, \ldots,$$

then $R - X$ is exponentially distributed.   (Oxford 1981F)

21. Let $X_1, X_2, \ldots, X_n$ be independent random variables each with characteristic function $\phi(t)$. Obtain the characteristic function of

$$Y_n = a_n + b_n(X_1 + X_2 + \cdots + X_n)$$

where a_n and b_n are arbitrary real numbers.

Suppose that $\phi(t) = e^{-|t|^\alpha}$, where $0 < \alpha \le 2$. Determine a_n and b_n such that Y_n has the same distribution as X_1 for $n = 1, 2, \ldots$. Find the probability density functions of X_1 when $\alpha = 1$ and when $\alpha = 2$.   (Oxford 1980F)

8

The two main limit theorems

8.1   The law of averages

We aim in this chapter to describe the two main limit theorems of probability theory, namely the 'law of large numbers' and the 'central limit theorem'. We begin with the law of large numbers.

Here is an example of the type of phenomenon which we are thinking about. Before writing this sentence we threw a fair die one million times (with the aid of a small computer, actually) and kept a record of the results. The average of the numbers which we threw was 3.500867; since the mean outcome of each throw is $\frac{1}{6}(1 + 2 + \cdots + 6) = 3\frac{1}{2}$, this number is not too surprising. If x_i is the result of the ith throw then most people would accept that the running averages

$$a_n = \frac{1}{n}(x_1 + x_2 + \cdots + x_n)$$

(1)

approach the mean value $3\frac{1}{2}$ as n gets larger and larger. Indeed, the foundations of probability theory are based upon our belief that sums of the form (1) converge to some limit as $n \to \infty$. It is upon the ideas of 'repeated experimentation' and 'the law of averages' that many of our notions of chance are founded. Accordingly, we should like to find a theorem of probability theory which says something like 'if we repeat an experiment many times, then the average of the results approaches the underlying mean value'.

With the above example about throwing a die in the backs of our minds, we suppose that we have a sequence $X_1, X_2, \ldots$ of independent and identically distributed random variables, each having mean value μ. We should like to prove that the average

$$\frac{1}{n}(X_1 + X_2 + \cdots + X_n)$$

(2)

converges as $n \to \infty$ to the underlying mean value μ. There are various ways in which random variables can be said to converge (advanced textbooks generally list four to six such ways). One very simple way is as follows. We say that the sequence $Z_1, Z_2, \ldots$ of random variables *converges in mean square to the (limit) random variable Z* if

(3)
$$E((Z_n - Z)^2) \to 0 \quad \text{as } n \to \infty.$$

If this holds, we write '$Z_n \to Z$ in mean square as $n \to \infty$'. Here is a word of motivation for this definition. Remember that if Y is a random variable and $E(Y^2) = 0$ then Y equals 0 with probability 1. If $E((Z_n - Z)^2) \to 0$ then it follows that $Z_n - Z$ tends to 0 (in some sense) as $n \to \infty$.

Example 4
Let Z_n be a discrete random variable with mass function

$$P(Z_n = 1) = \frac{1}{n}, \quad P(Z_n = 2) = 1 - \frac{1}{n}.$$

Then Z_n converges to the constant random variable 2 in mean square as $n \to \infty$, since

$$E((Z_n - 2)^2) = (1 - 2)^2 \frac{1}{n} + (2 - 2)^2 \left(1 - \frac{1}{n}\right)$$

$$= \frac{1}{n} \to 0 \text{ as } n \to \infty. \qquad \square$$

It is often quite simple to show convergence in mean square: just calculate a certain expectation and take the limit as $n \to \infty$. It is this type of convergence which appears in our first law of large numbers.

Theorem 8A
Mean-square law of large numbers. Let $X_1, X_2, \ldots$ be a sequence of independent random variables, each having mean value μ and variance σ^2. The average of the first n of the X's satisfies, as $n \to \infty$,

(5)
$$\frac{1}{n}(X_1 + X_2 + \cdots + X_n) \to \mu \text{ in mean square.}$$

Proof
This is a straightforward calculation. We write

$$S_n = X_1 + X_2 + \cdots + X_n$$

for the *n*th partial sum of the *X*'s. Then

$$E\left(\frac{1}{n} S_n\right) = \frac{1}{n} E(X_1 + X_2 + \cdots + X_n)$$

$$= \frac{1}{n} n\mu = \mu,$$

and so

$$\mathrm{E}\left(\left[\frac{1}{n}S_n - \mu\right]^2\right)^2 = \mathrm{var}\left(\frac{1}{n}S_n\right)$$

$$= n^{-2}\,\mathrm{var}(X_1 + X_2 + \cdots + X_n) \quad \text{by (7.10)}$$

$$= n^{-2}(\mathrm{var}\,X_1 + \cdots + \mathrm{var}\,X_n) \quad \begin{array}{l}\text{by independence}\\ \text{and (7.16)}\end{array}$$

$$= n^{-2}n\sigma^2$$

$$= n^{-1}\sigma^2 \to 0 \quad \text{as } n \to \infty.$$

Hence $n^{-1}S_n \to \mu$ in mean square as $n \to \infty$.   □

It is customary to assume that the random variables in the law of large numbers are identically distributed as well as independent. We demand here only that the X's have the same mean and variance.

Exercises   1. Let Z_n be a discrete random variable with mass function

$$\mathrm{P}(Z_n = n^\alpha) = \frac{1}{n}, \qquad \mathrm{P}(Z_n = 0) = 1 - \frac{1}{n}.$$

Show that Z_n converges to 0 in mean square if and only if $\alpha < \frac{1}{2}$.

2. Let $Z_1, Z_2, \ldots$ be a sequence of random variables which converges to the random variable Z in mean square. Show that $aZ_n + b \to aZ + b$ in mean square as $n \to \infty$, for any real numbers a and b.

3. Let N_n be the number of occurrences of 5 or 6 in n throws of a fair die. Use Theorem 8A to show that, as $n \to \infty$,

$$\frac{1}{n}N_n \to \tfrac{1}{3} \text{ in mean square.}$$

4. Show that the conclusion of Theorem 8A remains valid if the assumption that the X's are independent is replaced by the weaker assumption that they are uncorrelated.

8.2   Chebyshev's inequality and the weak law

The earliest versions of the law of large numbers were found in the eighteenth century and dealt with a form of convergence different from convergence in mean square. This other mode of convergence also has intuitive appeal and is defined in the following way. We say that the sequence $Z_1, Z_2, \ldots$ of random variables *converges in probability* to Z as $n \to \infty$ if

(6)   for all $\varepsilon > 0$, $\mathrm{P}(|Z_n - Z| > \varepsilon) \to 0$ as $n \to \infty$.

If this holds, we write '$Z_n \to Z$ in probability as $n \to \infty$'. Condition (6) requires that for all small positive δ and ε it is the case that $|Z_n - Z| \leq \varepsilon$ with probability at least $1 - \delta$, for all large values of n.

It is not clear at first sight how the two types of convergence (in mean square and in probability) are related to one another. It turns out that convergence in mean square is a more powerful property than convergence in probability, and we make this more precise in the next theorem.

Theorem 8B   *If $Z_1, Z_2, \ldots$ is a sequence of random variables and $Z_n \to Z$ in mean square as $n \to \infty$, then $Z_n \to Z$ in probability also.*

The proof of this follows immediately from a famous inequality which is usually ascribed to Chebyshev but which was discovered independently by Bienaymé and others. There are many forms of this inequality in the probability literature, and we feel that the following is the simplest.

Theorem 8C   *Chebyshev's inequality. If Y is a random variable and $\mathsf{E}(Y^2) < \infty$, then*

(7)
$$\mathsf{P}(|Y| \geq a) \leq \frac{1}{a^2} \mathsf{E}(Y^2) \qquad \text{for all } a > 0.$$

Proof   By the partition theorem for expectations (Theorem 2C deals with the case of discrete random variables, but it is clear that the corresponding result is valid for all random variables, regardless of their type)

(8)
$$\mathsf{E}(Y^2) = \mathsf{E}(Y^2 \mid A)\mathsf{P}(A) + \mathsf{E}(Y^2 \mid A^c)\mathsf{P}(A^c)$$

where $A = \{\omega \in \Omega : |Y(\omega)| \geq a\}$ is the event that $|Y| \geq a$. Each term on the right-hand side of (8) is non-negative, and so

$$\mathsf{E}(Y^2) \geq \mathsf{E}(Y^2 \mid A)\mathsf{P}(A)$$
$$\geq a^2 \mathsf{P}(A),$$

since if A occurs then $|Y| \geq a$ and $Y^2 \geq a^2$. $\qquad\square$

Proof of Theorem 8B   We apply Chebyshev's inequality to the random variable $Y = Z_n - Z$ to find that

$$\mathsf{P}(|Z_n - Z| > \varepsilon) \leq \frac{1}{\varepsilon^2} \mathsf{E}((Z_n - Z)^2) \qquad \text{for } \varepsilon > 0.$$

If $Z_n \to Z$ in mean square as $n \to \infty$, the right-hand side here tends to 0 as $n \to \infty$ and so the left-hand side tends to 0 for all $\varepsilon > 0$ as required. $\qquad\square$

The converse of Theorem 8B is false: there exist sequences of random variables which converge in probability but not in mean square, and there is an example at the end of this section.

The mean-square law of large numbers (Theorem 8A) combines with Theorem 8B to produce what is commonly called the 'weak law'.

Theorem 8D
Weak law of large numbers. Let $X_1, X_2, \ldots$ be a sequence of independent random variables, each having mean value μ and variance σ^2. The average of the first n of the X's satisfies, as $n \to \infty$,

(9)
$$\frac{1}{n}(X_1 + X_2 + \cdots + X_n) \to \mu \text{ in probability.}$$

The principal reasons for stating both the mean square law and the weak law are historical and traditional—the first laws of large numbers to be proved were in terms of convergence in probability. There is also a good mathematical reason for stating the weak law separately—unlike the mean square law, the conclusion of the weak law is valid without the assumption that the X's have finite variance so long as they all have the same distribution. This is harder to prove than the form of the weak law presented above, and we defer its proof until Section 8.4 and the treatment of characteristic functions therein.

There are many forms of the laws of large numbers in the literature, and each has a set of assumptions and a set of conclusions. Some are difficult to prove (with weak assumptions and strong conclusions) and others can be quite easy to prove (such as those above). Our selection is simple but contains many of the vital ideas. Incidentally, the weak law is called 'weak' because it may be formulated in terms of distributions alone; on the other hand, there is a more powerful 'strong law' which concerns intrinsically the convergence of random variables themselves.

Example 10
Here is an example of a sequence of random variables which converges in probability but not in mean square. Suppose that Z_n is a random variable with mass function

$$P(Z_n = 0) = 1 - \frac{1}{n}, \qquad P(Z_n = n) = \frac{1}{n}.$$

Then, for $\varepsilon > 0$ and all large n,

$$P(|Z_n| > \varepsilon) = P(Z_n = n)$$

$$= \frac{1}{n} \to 0 \text{ as } n \to \infty,$$

giving that $Z_n \to 0$ in probability. On the other hand

$$E((Z_n - 0)^2) = E(Z_n^2)$$

$$= 0 \cdot \left(1 - \frac{1}{n}\right) + n^2 \frac{1}{n}$$

$$= n \quad \to \infty \text{ as } n \to \infty,$$

so Z_n does not converge to 0 in mean square.   □

Exercises   5. Prove the following alternative form of Chebyshev's inequality: if X is a random variable with finite variance and $a > 0$, then

$$P(|X - E(X)| > a) \leq \frac{1}{a^2} \text{var}(X).$$

6. Use Chebyshev's inequality to show that the probability that in n throws of a fair die the number of sixes lies between $\frac{1}{6}n - \sqrt{n}$ and $\frac{1}{6}n + \sqrt{n}$ is at least $\frac{31}{36}$.

7. Show that if $Z_n \to Z$ in probability then, as $n \to \infty$,

$$aZ_n + b \to aZ + b \quad \text{in probability,}$$

for any real numbers a and b.

8.3   The central limit theorem

Our second main result is the central limit theorem. This also concerns sums of independent random variables. Let $X_1, X_2, \ldots$ be independent and identically distributed random variables, each with mean μ and non-zero variance σ^2. We know from the law of large numbers that the sum $S_n = X_1 + X_2 + \cdots + X_n$ is about as big as $n\mu$ for large n, and the next natural problem is to determine the order of the difference $S_n - n\mu$. It turns out that this difference has order $\sqrt{n}$.

Rather than work with the sum S_n directly, we work on the so-called *standardized* version of S_n,

(11)
$$Z_n = \frac{S_n - E(S_n)}{\sqrt{\text{var}(S_n)}};$$

this is a linear function $Z_n = a_n S_n + b_n$ of S_n where a_n and b_n have been chosen so that $E(Z_n) = 0$ and $\text{var}(Z_n) = 1$. Note that

$$E(S_n) = E(X_1) + E(X_2) + \cdots + E(X_n) \qquad \text{by (6.41)}$$

also,
$$= n\mu;$$

$$\text{var}(S_n) = \text{var}(X_1) + \cdots + \text{var}(X_n) \qquad \text{by independence and}$$
$$= n\sigma^2, \qquad\qquad\qquad\qquad\qquad\qquad (7.16)$$

(12)

and so

$$Z_n = \frac{S_n - n\mu}{\sigma\sqrt{n}}.$$

It is a remarkable fact that the distribution of Z_n settles down to a limit as $n \to \infty$; even more remarkable is the fact that the limiting distribution of Z_n is the normal distribution with mean 0 and variance 1, irrespective of the original distribution of the X's. This remarkable theorem is one of the most beautiful in mathematics and is known as the 'central limit theorem'.

Theorem 8E

Central limit theorem. Let $X_1, X_2, \ldots$ be independent and identically distributed random variables, each having mean value μ and non-zero variance σ^2. The standardized version

$$Z_n = \frac{S_n - n\mu}{\sigma\sqrt{n}}$$

of the sum $S_n = X_1 + X_2 + \cdots + X_n$ satisfies, as $n \to \infty$,

(13)

$$P(Z_n \le x) \to \int_{-\infty}^{x} \frac{1}{\sqrt{(2\pi)}} e^{-\frac{1}{2}u^2} du \qquad \textit{for } x \in \mathbb{R}.$$

The right-hand side of (13) is just the distribution function of the normal distribution with mean 0 and variance 1; thus (13) asserts that

$$P(Z_n \le x) \to P(Y \le x) \qquad \text{for } x \in \mathbb{R},$$

where Y is a random variable with this standard normal distribution.

Special cases of the central limit theorem were proved by de Moivre (in about 1733) and Laplace, who considered the case when the X's have the Bernoulli distribution. Liapounov proved the first general version in about 1901, but the details of his proof were very complicated. Here we shall give an elegant and short proof based on the method of moment generating functions. As one of our tools we shall use a special case of a fundamental theorem of analysis, and we present this next without proof. There is therefore a sense in which our 'short and elegant' proof is a fraud: it is only a partial proof, some of the analytical details being packaged elsewhere.

Theorem 8F

Continuity theorem. Let $Z_1, Z_2, \ldots$ be a sequence of random variables with moment generating functions $M_1, M_2, \ldots$ and suppose that, as $n \to \infty$,

(14)

$$M_n(t) \to e^{\frac{1}{2}t^2} \qquad \textit{for } t \in \mathbb{R}.$$

Then

$$P(Z_n \le x) \to \int_{-\infty}^x \frac{1}{\sqrt{(2\pi)}} e^{-\frac{1}{2}u^2} du \qquad \textit{for } x \in \mathbb{R}.$$

In other words, the distribution function of Z_n converges to the distribution function of the normal distribution if the moment generating function of Z_n converges to the moment generating function of the normal distribution. We shall use this to prove the central limit theorem in the case when the X's have a common moment generating function

$$M_X(t) = E(\exp(tX_i)) \qquad \text{for } i = 1, 2, \dots,$$

although we stress that the central limit theorem is valid even when this expectation does not exist so long as both the mean and the variance of the X's are finite.

Proof of Theorem 8E

Let

$$U_i = X_i - \mu.$$

Then $U_1, U_2, \dots$ are independent and identically distributed random variables with mean and variance given by

(15) $$E(U_i) = 0, \qquad E(U_i^2) = \text{var}(U_i) = \sigma^2,$$

and moment generating function

$$M_U(t) = M_X(t) e^{-\mu t}.$$

Now

$$Z_n = \frac{S_n - n\mu}{\sigma \sqrt{n}} = \frac{1}{\sigma \sqrt{n}} \sum_{i=1}^n U_i,$$

giving that Z_n has moment generating function

(16) $$M_n(t) = E(\exp(tZ_n))$$

$$= E\left(\exp\left(\frac{t}{\sigma \sqrt{n}} \sum_{i=1}^n U_i\right)\right),$$

$$= \left[M_U\left(\frac{t}{\sigma \sqrt{n}}\right)\right]^n \qquad \text{by (7.35) and (7.36).}$$

We need to know the behaviour of $M_U(t/(\sigma \sqrt{n}))$ for large n, and to this end we use Theorem 7F to expand $M_U(x)$ as a power series about $x = 0$:

$$M_U(x) = 1 + x E(U_1) + \tfrac{1}{2} x^2 E(U_1^2) + o(x^2)$$

$$= 1 + \tfrac{1}{2} \sigma^2 x^2 + o(x^2) \qquad \text{by (15).}$$

Substitute this into (16) with $x = t/(\sigma\sqrt{n})$ where t is fixed to obtain

$$M_n(t) = \left[1 + \frac{t^2}{2n} + o\left(\frac{1}{n}\right)\right]^n$$

$$\to \exp(\tfrac{1}{2}t^2) \quad \text{as } n \to \infty,$$

and the result follows from Theorem 8F. Note that this proof requires the existence of $M_X(t)$ for values of t near 0 only, and this is consistent with the discussion before Theorem 7D. $\qquad\square$

Exercises 8. A fair die is thrown 12,000 times. Use the central limit theorem to find values of a and b such that

$$P(1900 < S < 2200) \approx \int_a^b \frac{1}{\sqrt{(2\pi)}} e^{-\frac{1}{2}x^2}\, dx,$$

where S is the total number of sixes thrown.

9. For $n = 1, 2, \ldots$, let X_n be a random variable having the gamma distribution with parameters n and 1. Show that the moment generating function of $Z_n = n^{-\frac{1}{2}}(X_n - n)$ is

$$M_n(t) = e^{-t\sqrt{n}}\left(1 - \frac{t}{\sqrt{n}}\right)^{-n},$$

and deduce that, as $n \to \infty$,

$$P(Z_n \le x) \to \int_{-\infty}^x \frac{1}{\sqrt{(2\pi)}} e^{-\frac{1}{2}u^2}\, du \qquad \text{for all } x \in \mathbb{R}.$$

8.4 Convergence in distribution, and characteristic functions†

We have now encountered the ideas of convergence in mean square and convergence in probability, and we have seen that the former implies the latter. To these two types of convergence we are about to add a third. We motivate this by recalling the conclusion of the central limit theorem (Theorem 8E): the distribution function of a certain average Z_n converges as $n \to \infty$ to the distribution function of the normal distribution. This notion of the convergence of distribution functions may be set in a much more general context as follows. We say that the sequence $Z_1, Z_2, \ldots$ *converges in distribution to Z as $n \to \infty$* if

$$P(Z_n \le x) \to P(Z \le x) \qquad \text{for all } x \in C,$$

where C is the set of points of the real line $\mathbb{R}$ at which the distribution function $F_Z(z) = P(Z \le z)$ is continuous. This condition involving points of continuity is an unfortunate complication to the definition, but turns out to be desirable.

† Beginners to probability theory may wish to omit this section.

Convergence in distribution is a property of the distributions of random variables rather than a property of the random variables themselves, and for this reason explicit reference to the limit random variable Z is often omitted. For example, the conclusion of the central limit theorem may be expressed as 'Z_n converges in distribution to the normal distribution with mean 0 and variance 1'.

Theorem 8B asserts that convergence in mean square implies convergence in probability. It turns out that convergence in distribution is a weaker property than both of these.

Theorem 8F   *If $Z_1, Z_2, \ldots$ is a sequence of random variables and $Z_n \to Z$ in probability as $n \to \infty$, then $Z_n \to Z$ in distribution also.*

The converse assertion is generally false; see the forthcoming Example 18 for a sequence of random variables which converges in distribution but not in probability. The next theorem describes a partial converse.

Theorem 8G   *Let $Z_1, Z_2, \ldots$ be a sequence of random variables which converges in distribution to the constant c. Then Z_n converges to c in probability also.*

Proof of Theorem 8F   Suppose $Z_n \to Z$ in probability, and write

$$F_n(z) = \mathsf{P}(Z_n \le z), \qquad F(z) = \mathsf{P}(Z \le z)$$

for the distribution functions of Z_n and Z. Let $\varepsilon > 0$, and suppose that F is continuous at the point z. Then

$$
\begin{aligned}
F_n(z) &= \mathsf{P}(Z_n \le z) \\
&= \mathsf{P}(Z_n \le z, \, Z \le z + \varepsilon) + \mathsf{P}(Z_n \le z, \, Z > z + \varepsilon) \\
&\le \mathsf{P}(Z \le z + \varepsilon) + \mathsf{P}(Z - Z_n > \varepsilon) \\
&\le F(z + \varepsilon) + \mathsf{P}(|Z_n - Z| > \varepsilon).
\end{aligned}
$$

Similarly,

$$
\begin{aligned}
F(z - \varepsilon) &= \mathsf{P}(Z \le z - \varepsilon) \\
&= \mathsf{P}(Z \le z - \varepsilon, \, Z_n \le z) + \mathsf{P}(Z \le z - \varepsilon, \, Z_n > z) \\
&\le \mathsf{P}(Z_n \le z) + \mathsf{P}(Z_n - Z > \varepsilon) \\
&\le F_n(z) + \mathsf{P}(|Z_n - Z| > \varepsilon).
\end{aligned}
$$

Thus

(17) $$F(z - \varepsilon) - \mathsf{P}(|Z_n - Z| > \varepsilon) \le F_n(z) \le F(z + \varepsilon) + \mathsf{P}(|Z_n - Z| > \varepsilon).$$

We let $n \to \infty$ and $\varepsilon \downarrow 0$ throughout these inequalities. The left-hand side of (17) behaves as follows:

$$F(z - \varepsilon) - P(|Z_n - Z| > \varepsilon) \to F(z - \varepsilon) \qquad \text{as } n \to \infty$$
$$\to F(z) \qquad \text{as } \varepsilon \downarrow 0,$$

where we have used the facts that $Z_n \to Z$ in probability and that F is continuous at z, respectively. Similarly, the right-hand side of (17) satisfies

$$F(z + \varepsilon) + P(|Z_n - Z| > \varepsilon) \to F(z + \varepsilon) \qquad \text{as } n \to \infty$$
$$\to F(z) \qquad \text{as } \varepsilon \downarrow 0.$$

Thus the left- and right-hand sides of (17) have the same limit $F(z)$, implying that the central term $F_n(z)$ satisfies $F_n(z) \to F(z)$ as $n \to \infty$. Hence $Z_n \to Z$ in distribution. $\qquad \square$

Proof of Theorem 8G Suppose that $Z_n \to c$ in distribution as $n \to \infty$. It follows that the distribution function F_n of Z_n satisfies

$$F_n(z) \to \begin{cases} 0 & \text{if } z < c, \\ 1 & \text{if } z > c. \end{cases}$$

Thus, for $\varepsilon > 0$,

$$P(|Z_n - c| > \varepsilon) = P(Z_n < c - \varepsilon) + P(Z_n > c + \varepsilon)$$
$$\leq F_n(c - \varepsilon) + 1 - F_n(c + \varepsilon)$$
$$\to 0 + 1 - 1 = 0 \qquad \text{as } n \to \infty. \qquad \square$$

Here is an example of a sequence of random variables which converge in distribution but not in probability.

Example 18 Let U be a random variable which takes one of the values -1 or 1, each with probability $\frac{1}{2}$. We define the sequence $Z_1, Z_2, \ldots$ by

$$Z_n = \begin{cases} U & \text{if } n \text{ is odd}, \\ -U & \text{if } n \text{ is even}. \end{cases}$$

It is clear that $Z_n \to U$ in distribution as $n \to \infty$, since each variable in the sequence has the same distribution. On the other hand

$$Z_n - U = \begin{cases} 0 & \text{if } n \text{ is odd}, \\ -2U & \text{if } n \text{ is even}, \end{cases}$$

so that $P(|Z_{2m} - U| > 1) = P(|U| > \frac{1}{2}) = 1$ for all m; hence Z_n does not converge to U in probability. $\qquad \square$

Next we return to characteristic functions. In proving the central limit theorem we employed a result (Theorem 8F) linking the convergence of moment generating functions to convergence in distribution. This result is a weak form of the so-called continuity theorem, a much more powerful version of which we present next.

Theorem 8H

Continuity Theorem. Let $Z, Z_1, Z_2, \ldots$ be random variables with characteristic functions $\phi, \phi_1, \phi_2, \ldots$. Then $Z_n \to Z$ in distribution as $n \to \infty$ if and only if

$$\phi_n(t) \to \phi(t) \qquad \text{for all } t \in \mathbb{R}.$$

This is a difficult theorem to prove—see Feller (1971, p. 481). We close the section with several examples of this theorem in action.

Example 19

Suppose that $Z_n \to Z$ in distribution and $a, b \in \mathbb{R}$. Prove that $aZ_n + b \to aZ + b$ in distribution.

Solution

Let ϕ_n be the characteristic function of Z_n and ϕ the characteristic function of Z. By Theorem 8H, $\phi_n(t) \to \phi(t)$ as $n \to \infty$. The characteristic function of $aZ_n + b$ is

$$
\begin{aligned}
\phi_{aZ_n+b}(t) &= e^{itb}\phi_n(at) && \text{by Theorem 7H} \\
&\to e^{itb}\phi(at) && \text{as } n \to \infty \\
&= \phi_{aZ+b}(t),
\end{aligned}
$$

and the result follows by another appeal to Theorem 8H. A direct proof of this fact using distribution functions is messy when a is negative. $\square$

Example 20

The weak law. Here is another proof of the weak law of large numbers (Theorem 8D) for the case of identically distributed random variables. Let $X_1, X_2, \ldots$ be independent and identically distributed random variables with mean μ, and let

$$U_n = \frac{1}{n}(X_1 + X_2 + \cdots + X_n).$$

From Theorem 7H, the characteristic function ψ_n of U_n is given by

(21)
$$\psi_n(t) = \phi_X(t/n)^n,$$

where ϕ_X is the common characteristic function of the X's. By Theorem 7G

$$\phi_X(t) = 1 + it\mu + o(t).$$

Substitute this into (21) to obtain

$$\psi_n(t) = \left[1 + \frac{it\mu}{n} + o\left(\frac{t}{n}\right)\right]^n$$

$$\to e^{i\mu t} \quad \text{as } n \to \infty.$$

The limit here is the characteristic function of the constant μ, and thus the Continuity Theorem implies that $U_n \to \mu$ in distribution as $n \to \infty$. A glance at Theorem 8G confirms that the convergence takes place in probability also, and we have proved a version of the weak law of large numbers. This version differs from the earlier one in two regards—we have assumed that the X's are identically distributed but we have made no assumption that they have finite variance. $\square$

Example 22 *Central limit theorem.* Our proof of the central limit theorem in Section 8.3 was valid only for random variables which possess finite moment generating functions. Very much the same arguments go through using characteristic functions, and thus Theorem 8E is true as it stands. $\square$

Exercises 10. Let $X_1, X_2, \ldots$ be independent random variables each having the Cauchy distribution. Show that $U_n = n^{-1}(X_1 + X_2 + \cdots + X_n)$ converges in distribution to the Cauchy distribution as $n \to \infty$. Compare this with the conclusion of the weak law of large numbers.

8.5 Problems

1. Let $X_1, X_2, \ldots$ be independent random variables each having the uniform distribution on the interval $(0, a)$, and let $Z_n = \max\{X_1, X_2, \ldots, X_n\}$. Show that
   (i) $Z_n \to a$ in probability as $n \to \infty$,
   (ii) $\sqrt{Z_n} \to \sqrt{a}$ in probability as $n \to \infty$,
   (iii) if $U_n = n(1 - Z_n)$ and $a = 1$ then
   $$P(U_n \le x) \to \begin{cases} 1 - e^{-x} & \text{if } x > 0, \\ 0 & \text{otherwise,} \end{cases}$$
   so that the distribution of U_n converges to the exponential distribution as $n \to \infty$.
2. By applying the central limit theorem to a sequence of random variables with the Bernoulli distribution, or otherwise, prove the following result in analysis. If $0 < p = 1 - q < 1$ and $x > 0$, then
   $$\sum \binom{n}{k} p^k q^{n-k} \to 2 \int_0^x \frac{1}{\sqrt{(2\pi)}} e^{-\frac{1}{2}u^2} du \qquad \text{as } n \to \infty,$$
   where the summation is over all values of k satisfying $np - x\sqrt{(npq)} \le k \le np + x\sqrt{(npq)}$.
3. Let X_n be a discrete random variable with the binomial distribution, parameters n and p. Show that $n^{-1}X_n$ converges to p in probability as $n \to \infty$.
4. By applying the central limit theorem to a sequence of random

variables with the Poisson distribution, or otherwise, prove that

$$e^{-n}\left(1 + n + \frac{n^2}{2!} + \cdots + \frac{n^n}{n!}\right) \to \tfrac{1}{2} \qquad \text{as } n \to \infty.$$

5. Use the Cauchy–Schwarz inequality to prove that if $X_n \to X$ in mean square and $Y_n \to Y$ in mean square then $X_n + Y_n \to X + Y$ in mean square.

6. Use the Cauchy–Schwarz inequality to prove that if $X_n \to X$ in mean square then $E(X_n) \to E(X)$. Give an example of a sequence $X_1, X_2, \ldots$ such that $X_n \to X$ in probability but $E(X_n)$ does not converge to $E(X)$.

7. If $X_n \to X$ in probability and $Y_n \to Y$ in probability, show that $X_n + Y_n \to X + Y$ in probability.

8. Let $X_1, X_2, \ldots$ and $Y_1, Y_2, \ldots$ be independent random variables each having mean μ and variance σ^2 $(\neq 0)$. Show that

$$U_n = \frac{1}{\sqrt{(2n\sigma^2)}}\left(\sum_{i=1}^{n} X_i - \sum_{i=1}^{n} Y_i\right)$$

satisfies, as $n \to \infty$,

$$P(U_n \leq x) \to \int_{-\infty}^{x} \frac{1}{\sqrt{(2\pi)}} e^{-\frac{1}{2}u^2} \, du \qquad \text{for } x \in \mathbb{R}.$$

9. *Markov's inequality.* Adapt the proof of Chebyshev's inequality to show that if X is a random variable and $a > 0$ then

$$P(|X| \geq a) \leq \frac{1}{a} E(|X|).$$

10. Adapt the proof of Chebyshev's inequality to show that if X is a random variable and $a > 0$ then

$$P(|X| \geq a) \leq \frac{1}{g(a)} E(g(X)),$$

for any function $g : \mathbb{R} \to \mathbb{R}$ which satisfies
   (i) $g(x) = g(-x)$ for all $x \in \mathbb{R}$,
   (ii) $g(0) \geq 0$,
   (iii) g is strictly increasing on $[0, \infty)$.

11. Let X be a random variable which takes values in the interval $[-M, M]$ only. Show that

$$P(|X| \geq a) \geq \frac{E(|X|) - a}{M - a}$$

if $0 \leq a < M$.

12. Show that $X_n \to 0$ in probability if and only if

$$E\left(\frac{|X_n|}{1 + |X_n|}\right) \to 0 \qquad \text{as } n \to \infty.$$

13. Let (X_n) be a sequence of random variables which *converges in mean square*; state precisely what this means. Show that, for such a sequence (X_n),

$$E((X_n - X_m)^2) \to 0 \quad \text{as } m, n \to \infty.$$

If $E(X_n) = \mu$, $var(X_n) = \sigma^2$ for all n, show that the correlation between X_n and X_m satisfies $\rho(X_n, X_m) \to 1$ as $m, n \to \infty$.   (Bristol 1981)

14. Let the random variable Z have a standard normal distribution (i.e. $Z \sim N(0, 1)$). Evaluate $E(Z^2)$ and $E(Z^4)$. Let $Y = Z^2$. Find the probability density function for Y. Write down the mean and variance of Y. Now let $Z_1, Z_2, \ldots$ be a sequence of independent standard normal random variables. For each positive integer n define the random variable S_n by

$$S_n = \sum_{i=1}^n Z_i^2.$$

Use the central limit theorem to show that $P(S_n \leq n + k\sqrt{(2n)})$ tends to a limit as n tends to infinity for every constant k. Give a careful proof that $P(S_n \leq c)$ tends to zero as n tends to infinity for every constant c.
                (Bristol 1983)

*15. Let $X_1, X_2, \ldots$ be independent random variables each having distribution function F and density function f. The *order statistics* $X_{(1)}, X_{(2)}, \ldots, X_{(n)}$ of the finite subsequence $X_1, X_2, \ldots, X_n$ are obtained by rearranging the values of the X's in non-decreasing order. That is to say, $X_{(1)}$ is set to the smallest observed value of the X's, $X_{(2)}$ is set to the second smallest value, and so on, so that $X_{(n)} = \max\{X_1, X_2, \ldots, X_n\}$. The *sample median* Y_n of the sequence $X_1, X_2, \ldots, X_n$ is the 'middle value', so that Y_n is defined to be

$$Y_n = \begin{cases} X_{(m+1)} & \text{if } n = 2m + 1 \text{ is odd,} \\ \frac{1}{2}(X_{(m)} + X_{(m+1)}) & \text{if } n = 2m \text{ is even.} \end{cases}$$

Assume that $n = 2m + 1$ is odd and show that Y_n has density function

$$f_n(y) = (m + 1)\binom{n}{m}F(y)^m[1 - F(y)]^m f(y).$$

Deduce that if F has a unique median m then

$$P(Z_n \leq x) \to \int_{-\infty}^x \frac{1}{\sqrt{(2\pi)}} e^{-\frac{1}{2}u^2} du \qquad \text{for } u \in \mathbb{R},$$

where $Z_n = (Y_n - m)\sqrt{\{4n[f(m)]^2\}}$.

*16. The sequence (X_i) of independent, identically distributed random variables is such that

$$P(X_i = 0) = 1 - p, \qquad P(X_i = 1) = p.$$

If f is a continuous function on $[0, 1]$ prove that

$$B_n(p) = E\left(f\left(\frac{X_1 + \cdots + X_n}{n}\right)\right)$$

is a polynomial in p of degree at most n. Use Chebyshev's inequality to prove that for all p with $0 \leq p \leq 1$, and any $\varepsilon > 0$,

$$\sum_{k \in K} \binom{n}{k} p^k (1 - p)^{n-k} \leq \frac{1}{4n\varepsilon^2}.$$

where $K = \{k : 0 \leq k \leq n, |k/n - p| > \varepsilon\}$. Using this and the fact that f

must be bounded and uniformly continuous in $[0,1]$ prove the following version of Weierstrass's approximation theorem:

$$\lim_{n\to\infty} \sup_{0\le p\le 1} |f(p) - B_n(p)| = 0. \qquad \text{(Oxford 1976F)}$$

17. Let Z_n have the binomial distribution with parameters n and λ/n, where λ is fixed. Use characteristic functions to show that Z_n converges in distribution to the Poisson distribution, parameter λ, as $n\to\infty$.

18. Let Z_n have the geometric distribution with parameter λ/n, where λ is fixed. Show that Z_n/n converges in distribution as $n\to\infty$, and find the limiting distribution.

*19. Let $(X_k : k = 1, 2, \ldots)$ and $(Y_k : k = 1, 2, \ldots)$ be two sequences of independent random variables with

$$P(X_k = 1) = P(X_k = -1) = \frac{1}{2}$$

and

$$P(Y_k = 1) = P(Y_k = -1) = \frac{1}{2}\left(1 - \frac{1}{k^2}\right),$$

$$P(Y_k = k + 1) = P(Y_k = -k - 1) = \frac{1}{2k^2},$$

for $k = 1, 2, \ldots$. Let

$$S_n = \sum_{k=1}^{n} \frac{X_k}{\sqrt{n}}, \qquad T_n = \sum_{k=1}^{n} \frac{Y_k}{\sqrt{n}},$$

and let Z denote a normally distributed random variable with mean 0 and variance 1.

Prove or disprove the following:
  (i) S_n converges in distribution to Z,
  (ii) the mean and variance of T_n converge to the mean and variance of Z,
  (iii) T_n converges in distribution to Z.

State carefully any theorems which you use.   (Oxford 1980F)

*20. Let X_j, $j = 1, 2, \ldots, n$ be independent identically distributed random variables with probability density function $e^{-\frac{1}{2}x^2}/\sqrt{(2\pi)}$, $-\infty < x < \infty$. Show that the characteristic function of $Y = X_1^2 + \cdots + X_n^2$ is $[1 - 2i\theta]^{-\frac{1}{2}n}$. Consider a sequence of independent trials where the probability of success is p for each trial. Let N be the number of trials required to obtain a fixed number of k successes. Show that as p tends to zero the distribution of $2Np$ tends to the distribution of Y with $n = 2k$.   (Oxford 1979F)

21. Let $X_1, X_2, \ldots, X_n$ be independent and identically distributed random variables such that

$$P(X_1 = 1) = P(X_1 = -1) = \tfrac{1}{2}.$$

Derive the characteristic function of the random variable $Y_n = \sum_{j=1}^{n} a_j X_j$, where $a_1, a_2, \ldots, a_n$ are constants. In the special case $a_j = 2^{-j}$ for all j, show that as n tends to infinity, the distribution of Y_n approaches the uniform distribution on the interval $(-1, 1)$.

Any theorems which are needed about characteristic functions should be stated carefully but need not be proved.   (Oxford 1981F)

*22. X and Y are independent, identically distributed random variables with mean 0, variance 1, and characteristic function ϕ. If $X + Y$ and $X - Y$ are independent, then prove that

$$\phi(2t) = [\phi(t)]^3\phi(-t).$$

By making the substitution $\gamma(t) = \phi(t)/\phi(-t)$ or otherwise show that for any positive integer n,

$$\phi(t) = \left\{1 - \tfrac{1}{2}\left(\frac{t}{2^n}\right)^2 + o\left[\left(\frac{t}{2^n}\right)^2\right]\right\}^{4^n}.$$

Hence find the common distribution of X and Y. (Oxford 1976F)

23. Let $u(t)$ and $v(t)$ be the real and imaginary parts respectively of the characteristic function of the random variable X. Prove that

(a) $\qquad\qquad E(\cos^2 tX) = \tfrac{1}{2}[1 + u(2t)],$

(b) $\qquad\qquad E(\cos sX \cos tX) = \tfrac{1}{2}[u(s + t) + u(s - t)].$

Hence find the variance of $\cos tX$ and the covariance of $\cos tX$ and $\cos sX$ in terms of u and v.

Consider the special case when X is uniformly distributed on $[0, 1]$. Are the random variables $\{\cos j\pi X : j = 1, 2, \ldots\}$ (i) uncorrelated, (ii) independent? Justify your answers. (Oxford 1975F)

C. Random Processes

9
Branching processes

9.1 Random processes

Until now we have been developing the basic terminology and results of probability theory; next, we turn our attention to simple applications. The passing of time plays an essential part in the world which we inhabit, and consequently most applications of probability involve quantities which develop randomly as time passes. Such randomly evolving processes are called *random processes* or *stochastic processes,* and there are many different types of these. Most real processes in nature, such as the pollen count at Lords or the position of Swansea City in the football league, develop according to rules which are too complicated to describe exactly, and good probabilistic models for these processes can be very complicated indeed. We shall stick to some of the simplest random processes, and the specific processes which we shall consider in some depth are

(i) *branching processes*: modelling the growth of a self-reproducing population (such as mankind),

(ii) *random walks*: modelling the movement of a particle which moves erratically within the medium which contains it (a dust particle in the atmosphere, say),

(iii) *Poisson processes and related processes*: modelling processes such as the emission of radioactive particles from a slowly decaying source, or the length of the queue at the supermarket cash register.

9.2 A model for population growth

We define the term *nomad* to be a type of hypothetical object which is able to reproduce itself according to the following rules. At time $n = 0$ there exists a single nomad. This nomad lives for a unit of time and then, at time $n = 1$, it dies in the act of childbirth and is replaced by a family of offspring nomads. These nomads have similar biographies, each surviving only until time $n = 2$ and then each dying and being replaced by a family of offspring. This death-birth process continues at all subsequent time points $n = 3, 4, \ldots$. If we know the sizes of all the individual nomad families then we know everything

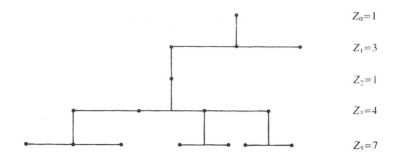

$Z_0 = 1$

$Z_1 = 3$

$Z_2 = 1$

$Z_3 = 4$

$Z_5 = 7$

Fig. 9.1 A typical nomad family tree, with generation sizes 1, 3, 1, 4, 7, . . .

about the development of the nomad population, and we might represent this in the usual way as a family tree (see Fig. 9.1). The problem is that different nomads may have different numbers of offspring, and these numbers may not be entirely predictable in advance. We shall assume here that the family sizes are random variables which satisfy the following two conditions:

(1)   the family sizes are independent random variables each taking values in $\{0, 1, 2, \ldots\}$,

(2)   the family sizes are identically distributed random variables with known mass function p, so that the number C of children of a typical nomad has mass function $P(C = k) = p(k)$ for $k = 0, 1, 2, \ldots$.

Such a process is called a *branching process* and may be used as a simple model for bacterial growth or the spread of a family name (to give but two examples). Having established the model, the basic problem is to say something about how the development of the process depends on the family-size mass function p. In order to avoid trivialities, we shall suppose throughout that $p(k) \neq 1$ for $k = 0, 1, 2, \ldots$.

Next we introduce some notation. The set of nomads born at time n is called the nth *generation* of the branching process, and we write Z_n for the number of such nomads. The evolution of the process is described by the sequence $Z_0, Z_1, \ldots$ of random variables and it is with this sequence that we work. Specific properties of the Z's are given in the next section, and we close this section with a list of interesting questions.

 (i) What is the mean and variance of Z_n?
 (ii) What is the mass function of Z_n?
(iii) What is the probability that nomadkind is extinct by time n?
(iv) What is the probability that nomadkind ultimately becomes extinct?

9.3 The generating-function method

The first step in the study of this branching process is to explain how to find the distributions of the Z's in terms of the family-size mass function p. Clearly $Z_0 = 1$ and

(3)
$$P(Z_1 = k) = p(k) \quad \text{for } k = 0, 1, 2, \ldots,$$

since Z_1 is the number of children of the founding nomad. It is not so easy to write down the mass function of Z_2 directly since Z_2 is the sum of a random number Z_1 of random family-sizes: writing C_i for the number of children of the ith nomad in the first generation, we have that

$$Z_2 = C_1 + C_2 + \cdots + C_{Z_1},$$

the sum of the family sizes of the Z_1 nomads in the first generation. More generally, for $n = 1, 2, \ldots,$

(4)
$$Z_n = C_1' + C_2' + \cdots + C_{Z_{n-1}}'$$

where $C_1', C_2', \ldots$ are the numbers of children of the nomads in the $(n-1)$th generation. The sum of a random number of random variables is treated better by using probability generating functions than by using mass functions, and thus we turn our attention to the probability generating functions of $Z_0, Z_1, \ldots$. We write

$$G_n(s) = E(s^{Z_n}) = \sum_{k=0}^{\infty} s^k P(Z_n = k)$$

for the probability generating function of Z_n, and

$$G(s) = \sum_{k=0}^{\infty} s^k p(k)$$

for the probability generating function of a typical family size. We wish to express G_n in terms of G and do this in the following theorem.

Theorem 9A *The probability generating functions $G, G_0, G_1, \ldots$ satisfy*

(5)
$$G_0(s) = s, \quad G_n(s) = G_{n-1}(G(s)), \quad \text{for } n = 1, 2, \ldots,$$

and hence G_n is the nth iterate of G:

(6)
$$G_n(s) = G(G(\cdots G(s) \cdots)) \quad \text{for } n = 0, 1, 2, \ldots.$$

Proof $Z_0 = 1$ and so $G_0(s) = s$. Equation (4) expresses Z_n as the sum of Z_{n-1} independent random variables each having probability generating function G, and so Theorem 4D may be applied with $X_i = C_i'$ and $N = Z_{n-1}$ to deduce that

(7)
$$G_n(s) = G_{n-1}(G(s)).$$

Hence
$$G_n(s) = G_{n-1}(G(s))$$
$$= G_{n-2}(G(G(s)))$$
$$= \cdots$$
$$= G_1(G(G(G(\cdots(s)\cdots))))$$

where $G_1 = G$ by (3). □

Theorem 9A contains all the information necessary for studying the development of the process. The next result is an immediate corollary.

Theorem 9B *The mean value of Z_n is*

(8)
$$\mathsf{E}(Z_n) = \mu^n$$

where $\mu = \sum_k kp(k)$ is the mean of the family-size distribution.

Proof By the elementary theory of probability generating functions,

$$\mathsf{E}(Z_n) = G_n'(1) \qquad \text{by (4.18)}$$
$$= G_{n-1}'(G(1))G'(1) \qquad \text{by (5)}$$
$$= G_{n-1}'(1)G'(1) \qquad \text{by (4.6)}$$
$$= \mathsf{E}(Z_{n-1})\mu.$$

Hence
$$\mathsf{E}(Z_n) = \mu\mathsf{E}(Z_{n-1})$$
$$= \mu^2\mathsf{E}(Z_{n-2})$$
$$= \cdots$$
$$= \mu^n\mathsf{E}(Z_0) = \mu^n. \qquad \square$$

The variance of Z_n may be derived similarly, in terms of the mean μ and the variance σ^2 of the family-size distribution.

It follows from Theorem 9B that

$$\mathsf{E}(Z_n) \to \begin{cases} 0 & \text{if } \mu < 1, \\ 1 & \text{if } \mu = 1, \\ \infty & \text{if } \mu > 1, \end{cases}$$

indicating that the behaviour of the process depends substantially on which of the three cases $\mu < 1$, $\mu = 1$, $\mu > 1$ holds. We shall see this in more detail in the next two sections, where it is shown that if $\mu \leq 1$ then the nomad population is bound to become extinct, whereas if $\mu > 1$ then there is a strictly positive probability that the line of descent of nomads will continue for ever. This dependence on the mean family size μ is quite natural since '$\mu < 1$' means that each nomad gives birth to (on average) strictly fewer nomads than are necessary to fill the gap caused by its death, whereas '$\mu > 1$' means that each death results (on average) in an increase in the population. The case when $\mu = 1$ is called *critical* since then the mean population size equals 1 for all time; in this case, random fluctuations ensure that the population size will take the value 0 sooner or later, and henceforth nomadkind will be extinct.

Exercises   1. Show that, in the above branching process,

$$G_n(s) = G_r(G_{n-r}(s))$$

for any $r = 0, 1, 2, \ldots, n$. This may be proved either directly from the conclusion of Theorem 9A or by adapting the method of proof of (7).

2. Suppose that each family-size of a branching process contains either one member only (with probability p) or is empty (with probability $1 - p$). Find the probability that the process becomes extinct at or before the nth generation.

3. Let μ and σ^2 be the mean and variance of the family-size distribution. Adapt the proof of Theorem 9B to show that the variance of Z_n, the size of the nth generation of the branching process, is given by

$$\text{var}(Z_n) = \begin{cases} n\sigma^2 & \text{if } \mu = 1, \\ \sigma^2 \mu^{n-1} \dfrac{\mu^n - 1}{\mu - 1} & \text{if } \mu \neq 1. \end{cases}$$

9.4   An example

The key to the analysis of branching processes is the functional equation

(9) $$G_n(s) = G_{n-1}(G(s)),$$

relating the probability generating functions of Z_n and Z_{n-1} and derived in Theorem 9A. There are a few instances in which this equation may be solved in closed form, and we consider one of these cases here. Specifically, we suppose that the mass function of each family size is given by

$$p(k) = (\tfrac{1}{2})^{k+1} \qquad \text{for } k = 0, 1, 2, \ldots,$$

so that each family size is one member smaller than a geometrically distributed random variable with parameter $\frac{1}{2}$ (remember (2.9)) and has probability generating function

$$G(s) = \sum_{k=0}^{\infty} s^k (\tfrac{1}{2})^{k+1} = \frac{1}{2-s} \qquad \text{for } |s| < 2.$$

We proceed as follows in order to solve (9). First, if $|s| \le 1$,

$$G_1(s) = G(s) = \frac{1}{2-s}.$$

Secondly, we apply (9) with $n = 2$ to find that

$$G_2(s) = G(G(s))$$

$$= \frac{1}{2 - (2-s)^{-1}} = \frac{2-s}{3-2s} \qquad \text{if } |s| \le 1.$$

The next step gives

$$G_3(s) = G_2(G(s)) = \frac{3-2s}{4-3s} \qquad \text{if } |s| \le 1;$$

it is natural to guess that

(10) $$G_n(s) = \frac{n - (n-1)s}{n+1-ns} \qquad \text{if } |s| \le 1,$$

for any $n \ge 1$, and this is proved easily from (9), by the method of induction. The mass function of Z_n follows by expanding the right-hand side of (10) as a power series in s, to find that the coefficient of s^k is

(11) $$P(Z_n = k) = \begin{cases} \dfrac{n}{n+1} & \text{if } k = 0, \\[2ex] \dfrac{n^{k-1}}{(n+1)^{k+1}} & \text{if } k = 1, 2, \dots. \end{cases}$$

In particular

$$P(Z_n = 0) = \frac{n}{n+1} \to 1 \quad \text{as } n \to \infty,$$

so that this branching process becomes extinct sooner or later, with probability 1.

There is a more general case which is of greater interest. Suppose that the mass function of each family size is given by

$$p(k) = pq^k \qquad \text{for } k = 0, 1, 2, \dots$$

where $0 < p = 1 - q < 1$; the previous example is the case when

$p = q = \frac{1}{2}$, but we suppose here that $p \neq \frac{1}{2}$ so that $p \neq q$. In this case

$$G(s) = \frac{p}{1 - qs} \qquad \text{if } |s| < q^{-1},$$

and the solution to (9) is

(12)
$$G_n(s) = p \frac{(q^n - p^n) - qs(q^{n-1} - p^{n-1})}{(q^{n+1} - p^{n+1}) - qs(q^n - p^n)},$$

valid for $n = 1, 2, \ldots$ and $|s| \leq 1$; again, this can be proved from (9) by induction on n. The mass function of Z_n is rather more complicated than (11) but may be expressed in very much the same way. The probability of extinction is found to be

$$P(Z_n = 0) = G_n(0) \qquad\qquad \text{by (4.6)}$$

$$= p \frac{q^n - p^n}{q^{n+1} - p^{n+1}}$$

$$= \frac{\mu^n - 1}{\mu^{n+1} - 1}$$

where $\mu = q/p$ is the mean family size. Hence

$$P(Z_n = 0) \to \begin{cases} 1 & \text{if } \mu < 1, \\ \mu^{-1} & \text{if } \mu > 1, \end{cases}$$

giving that ultimate extinction is certain if $\mu < 1$ and less than certain if $\mu > 1$. Combined with the result when $p = q = \frac{1}{2}$ and $\mu = q/p = 1$, this shows that ultimate extinction is certain if and only if $\mu \leq 1$. We shall see in the next section that this is a special case of a general result.

Exercises   4. Find the mean and variance of Z_n when the family-size distribution is given by $p(k) = pq^k$ for $k = 0, 1, 2, \ldots$, and $0 < p = 1 - q < 1$. Deduce that $\mathrm{var}(Z_n) \to 0$ if and only if $p > \frac{1}{2}$.

9.5   The probability of extinction

In the previous example, ultimate extinction of the branching process is certain if and only if the mean family-size μ satisfies $\mu \leq 1$. This conclusion is valid for *all* branching processes (except for the non-random branching process in which every family size equals 1 always) and we shall prove this. It is clear that if $Z_n = 0$ then $Z_m = 0$ for all $m \geq n$. The probability that the branching process is extinct (in that nomadkind has died out) by the nth generation is

(13)
$$e_n = P(Z_n = 0);$$

also
$$e_n \leq e_{n+1}$$
since '$Z_n = 0$' implies '$Z_{n+1} = 0$'. Hence
$$e = \lim_{n \to \infty} e_n$$
exists, and we call e the *probability of ultimate extinction* of the process†. How do we calculate e? Clearly if $p(0) = 0$ then $e = 0$, since all families are non-empty; the next theorem deals with the general case.

Theorem 9C *The probability e of ultimate extinction is the smallest non-negative root of the equation*

(14)
$$x = G(x).$$

Proof We note first that $e_n = G_n(0)$; this follows from (13) and (4.6). Now, from (5) and (6),
$$G_n(s) = G_{n-1}(G(s))$$
$$= G(G(\cdots(s)\cdots))$$
$$= G(G_{n-1}(s)).$$

Set $s = 0$ to find that $e_n = G_n(0)$ satisfies

(15)
$$e_n = G(e_{n-1}) \qquad \text{for } n = 1, 2, \ldots$$

with the boundary condition $e_0 = 0$. Let $n \to \infty$ here and use the fact that $e_n \to e$ to find that e satisfies the equation
$$e = G(e)$$
as required (we are using the fact that G is a power series with radius of convergence at least 1, giving that G is continuous on $[0, 1]$). To show that e is the smallest non-negative root of (14), suppose that η is any non-negative root of (14); we shall show that $e \leq \eta$. First, G is non-decreasing on $[0, 1]$ since it has non-negative coefficients, and hence
$$e_1 = G(0) \leq G(\eta) = \eta$$
by (15). Similarly,
$$e_2 = G(e_1) \leq G(\eta) = \eta$$
by (15), giving by induction that
$$e_n \leq \eta \qquad \text{for } n = 1, 2, \ldots.$$

Hence $e = \lim_{n \to \infty} e_n \leq \eta$. □

† We have used the continuity of probability measures (Theorem 1C) surreptitiously here.

The last theorem tells us how to find the probability of ultimate extinction. The next theorem tells us when extinction is *bound* to occur.

Theorem   *The probability e of ultimate extinction satisfies e = 1 if and only if the*
9D   *mean family-size μ satisfies μ ≤ 1.*

Recall from Section 9.2 that we have ruled out the special case with $p(1) = 1$ and $p(i) = 0$ if $i \neq 1$; all families have size 1 in this special case, so that $\mu = 1$ but $e = 0$.

Proof   We may suppose that $p(0) > 0$, since otherwise $e = 0$ and $\mu > 1$. We have seen that e is the smallest non-negative root of the equation $x = G(x)$. Now, on $[0, 1]$, G is
  (i) continuous (since its radius of convergence is at least 1),
  (ii) non-decreasing (since $G'(x) = \sum_k kp(k)x^{k-1} \geq 0$),
  (iii) convex (since $G''(x) = \sum_k k(k-1)p(k)x^{k-2} \geq 0$),
and so a sketch of G looks something like the curves in Fig. 9.2. Generally speaking, there are either one or two intersections between the curve $y = G(x)$ and the line $y = x$ in the interval $[0, 1]$. If the derivative $G'(1)$ of $G(x)$ at $x = 1$ satisfies $G'(1) > 1$ then the geometry of Fig. 9.2(a) implies that there are two distinct intersections (so that $e < 1$), whereas if $G'(1) \leq 1$ then Fig. 9.2(b) shows that there is a unique intersection, and $x = 1$ is the only root in $[0, 1]$ of the equation $x = G(x)$. But $G'(1) = \mu$, and so $e = 1$ if and only if $\mu \leq 1$.   □

Exercises   5. If the family-size distribution of a branching process has mass function $p(k) = pq^k$ for $k = 0, 1, 2, \ldots$ and $0 < p = 1 - q < 1$, use Theorem 9C to

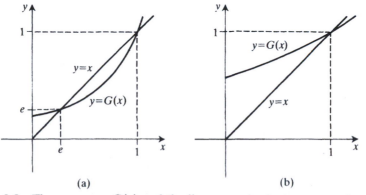

(a)   (b)

Fig. 9.2   The curve $y = G(x)$ and the line $y = x$, in the two cases when (a) $G'(1) > 1$ and (b) $G'(1) \leq 1$

show that the probability that the process becomes extinct ultimately is p/q if $p \leq \frac{1}{2}$.

6. If each family size of a branching process has the binomial distribution with parameters 2 and p $(=1-q)$ show that the probability of ultimate extinction is

$$e = \begin{cases} 1 & \text{if } 0 \leq p \leq \frac{1}{2}, \\ (q/p)^2 & \text{if } \frac{1}{2} \leq p \leq 1. \end{cases}$$

9.6 Problems

1. Let $X_1, X_2, \ldots$ be independent random variables each having mean μ and variance σ^2, and let N be a random variable which takes values in the positive integers and which is independent of the X's. Show that the sum

$$S = X_1 + X_2 + \cdots + X_N$$

has variance given by

$$\text{var}(S) = \sigma^2 \text{E}(N) + \mu^2 \text{var}(N).$$

If $Z_0, Z_1, \ldots$ are the generation sizes of a branching process in which each family size has mean μ and variance σ^2, use the above fact to show that

$$\text{var}(Z_n) = \sigma^2 \mu^{n-1} + \mu^2 \text{var}(Z_{n-1}),$$
$$= \mu \text{var}(Z_{n-1}) + \sigma^2 \mu^{2n-2}.$$

Deduce an expression for $\text{var}(Z_n)$ in terms of μ, σ^2 and n.

2. Use the result of Problem 1 to show that, if $Z_0, Z_1, \ldots$ is a branching process whose family sizes have mean μ (>1) and variance σ^2, then $\text{var}(Z_n/\mu^n) \to \sigma^2[\mu(\mu-1)]^{-1}$ as $n \to \infty$.

3. By using the partition theorem and conditioning on the value of Z_m, show that if $Z_0, Z_1, \ldots$ is a branching process with mean family size μ then

$$\text{E}(Z_m Z_n) = \mu^{n-m} \text{E}(Z_m^2) \qquad \text{if } m < n.$$

4. If $(Z_n : 0 \leq n < \infty)$ is a branching process in which $Z_0 = 1$ and the size of the rth generation Z_r has the generating function $P_r(s)$, prove that

$$P_n(s) = P_r(P_{n-r}(s)) \qquad \text{for } 1 \leq r \leq n-1.$$

Suppose that the process is modified so that the initial generation Z_0 is Poisson with parameter λ and the number of offspring of each individual is also Poisson with parameter λ. Find a function f such that if $H_n(s)$ is the generating function of the total number of individuals $Z_0 + Z_1 + \cdots + Z_n$ then

$$H_n(s) = f(sH_{n-1}(s)). \tag{Oxford 1977F}$$

5. A simple branching process $(X_n : n \geq 0)$ has $\text{P}(X_0 = 1) = 1$. Let the total number of individuals in the first n generations of the process be Z_n, with probability generating function Q_n. Prove that, for $n \geq 2$,

$$Q_n(s) = s P_1(Q_{n-1}(s)),$$

where P_1 is the probability generating function of the family-size distribution.

(Oxford 1974F)

10

Random walks

10.1 One-dimensional random walks

There are many practical instances of random walks. Many processes in physics involve atomic and sub-atomic particles which migrate about the space which they inhabit, and we may often model such motions by random-walk processes. In addition, random walks may often be detected in non-physical disguises, such as in models for gambling, epidemic spread, and stockmarket indices. We shall consider only the simplest type of random walk here, and we describe this in terms of a hypothetical particle which inhabits a one-dimensional set (that is to say, a line) and which moves randomly within this set as time passes.

For simplicity, we assume that both space and time are *discrete*; that is to say, we shall observe the particle's position at each of the time-points $0, 1, 2, \ldots$ (rather than at each time in $[0, \infty)$), and we assume that at each of these time-points the particle is located at one of the integer positions $\ldots, -2, -1, 0, 1, 2, \ldots$ of the real line (rather than at some point in the larger set $\mathbb{R}$). The particle moves in the following way. Let p be a real number satisfying $0 < p < 1$, and let $q = 1 - p$. During each unit of time the particle moves its location either one unit leftwards (with probability q) or one unit rightwards (with probability p). More specifically, if S_n denotes the location of the particle at time n then

$$S_{n+1} = \begin{cases} S_n + 1 & \text{with probability } p, \\ S_n - 1 & \text{with probability } q, \end{cases}$$

and we suppose that the random direction of each jump is independent of all earlier jumps. Therefore

(1) $$S_n = S_0 + X_1 + X_2 + \cdots + X_n \qquad \text{for } n = 0, 1, 2, \ldots,$$

where S_0 is the starting position and $X_1, X_2, \ldots$ are independent random variables each taking either the value -1 with probability q or the value 1 with probability p. We call the process $S_0, S_1, \ldots$ *simple random walk*; it is called *symmetric* random walk if $p = q = \frac{1}{2}$ and *asymmetric* random walk otherwise.

Example 2   *Gambling.* This example of a random walk is well known. A gambler enters a casino with £1000 in his pocket and sits at a table where he proceeds to play the following game. The croupier flips a coin repeatedly, and on each flip the coin shows heads with probability p and tails with probability $q = 1 - p$. Whenever the coin shows heads the casino pays the gambler £1, and whenever the coin shows tails the gambler pays the casino £1; that is, at each stage the gambler's capital increases by £1 with probability p and decreases by £1 with probability q. Writing S_n for the gambler's capital after n flips of the coin, we have that $S_0 = 1000$, and $S_0, S_1, \ldots$ is a simple random walk. If the casino refuses to extend credit, then the gambler becomes bankrupt at the first time the random walk visits the point 0, and he may be interested in the probability that he ultimately becomes bankrupt. It turns out that

(3)
$$P(S_n = 0 \text{ for some } n \ge 1) \begin{cases} <1 & \text{if } p > \frac{1}{2}, \\ =1 & \text{if } p \le \frac{1}{2}, \end{cases}$$

so that a compulsive gambler can avoid ultimate bankruptcy (with positive probability) if and only if the odds are stacked in his favour. We shall respect the time-honoured tradition of probability textbooks by returning to this example later. $\square$

Exercises   In these exercises $S_0, S_1, \ldots$ is a random walk on the integers in which p $(=1-q)$ is the probability that any given step is to the right.
1. Find the mean and variance of S_n when $S_0 = 0$.
2. Find the probability that $S_n = n + k$ given that $S_0 = k$.

10.2   Transition probabilities

Consider a simple random walk starting from the point $S_0 = 0$. Given a time-point n and a location k, what is the probability that $S_n = k$? The probabilities of such transitions of the random walker are calculated by primitive arguments involving counting, and the following result is typical.

Theorem 10A   *Let $u_n = P(S_n = S_0)$ be the probability that a simple random walk revisits its starting point at time n. Then $u_n = 0$ if n is odd, and*

(4)
$$u_{2m} = \binom{2m}{m} p^m q^m$$

if $n = 2m$ is even.

Proof   We may suppose without loss of generality that $S_0 = 0$. If $S_0 = S_n = 0$, then the walk made equal numbers of leftward and rightward steps in its first n steps; this is impossible if n is odd, giving that $u_n = 0$ if n is odd. Suppose now that $n = 2m$ for some integer m. From (1),

$$S_{2m} = X_1 + X_2 + \cdots + X_{2m},$$

so that $S_{2m} = 0$ if and only if exactly m of the X's equal $+1$ and exactly m equal -1 (giving m rightward steps and m leftward steps). There are $\binom{2m}{m}$ ways of dividing the X's into two sets with equal sizes, and the probability that each of the first set equals $+1$ and each of the second set equals -1 is $p^m q^m$. Hence

$$u_{2m} = \binom{2m}{m} p^m q^m$$

as required.   □

More general transition probabilities may be calculated similarly. Perhaps the simplest argument proceeds as follows. For $i \geq 1$, the random variable $\frac{1}{2}(X_i + 1)$ has the Bernoulli distribution with parameter p, giving that $B_n = \frac{1}{2}(S_n + n)$ has the binomial distribution with parameters n and p. Hence

(5)
$$P(S_n = k \mid S_0 = 0) = P(B_n = \tfrac{1}{2}(n + k))$$

$$= \binom{n}{\frac{1}{2}(n + k)} p^{\frac{1}{2}(n+k)} q^{\frac{1}{2}(n-k)}$$

whenever k is such that $\frac{1}{2}(n + k)$ is an integer between 0 and n. The result of Theorem 10A is retrieved by setting $k = 0$.

Exercises   3. Find $P(S_{2n+1} = 1 \mid S_0 = 0)$.
4. Show that

$$\sum_{n=0}^{\infty} u_n \begin{cases} < \infty & \text{if } p \neq q, \\ = \infty & \text{if } p = q, \end{cases}$$

and deduce that an asymmetric random walk revisits its starting point only finitely often with probability 1. You will need Stirling's formula: $n! \approx n^{n+\frac{1}{2}} e^{-n} \sqrt{(2\pi)}$ for large values of n.
5. Consider a two-dimensional random walk in which a particle moves between the points $\{(i, j) : i, j = \ldots, -1, 0, 1, \ldots\}$ with integral coordinates in the plane. Let p, q, r, s be numbers such that $0 < p, q, r, s < 1$ and $p + q + r + s = 1$. If the particle is at position (i, j) at time n then its position at time $n + 1$ is

$(i + 1, j)$ with probability p, $(i, j + 1)$ with probability q,
$(i - 1, j)$ with probability r, $(i, j - 1)$ with probability s,

and successive moves are independent of each other. Writing S_n for the position of the particle after n moves, we have that

$$S_{n+1} = \begin{cases} S_n + (1, 0) & \text{with probability } p, \\ S_n + (0, 1) & \text{with probability } q, \\ S_n - (1, 0) & \text{with probability } r, \\ S_n - (0, 1) & \text{with probability } s, \end{cases}$$

and we suppose that $S_0 = (0, 0)$. Let v_n be the probability that the particle revisits its starting point at time n, so that $v_n = P(S_n = (0, 0))$. Show that $v_n = 0$ if n is odd and

$$v_{2m} = \sum_{k=0}^{m} \frac{(2m)!}{k!^2(m-k)!^2} (pr)^k (qs)^{m-k}$$

if $n = 2m$ is even.

10.3 Recurrence and transience in random walks

Consider a simple random walk starting from the point $S_0 = 0$. In the subsequent motion, the random walk may or may not revisit its starting point. If the walk is bound (with probability 1) to revisit its starting point then we call it *recurrent*; otherwise we call it *transient*. We shall see that a simple random walk is recurrent if and only if it is symmetric (in that $p = q = \frac{1}{2}$), and there is a simple intuitive reason why this is the case. The position at time n is the sum $S_n = X_1 + X_2 + \cdots + X_n$ of independent random variables, each having mean value $E(X) = p - q$ and finite variance, and hence

$$\frac{1}{n} S_n \to p - q \qquad \text{as } n \to \infty$$

in mean square, by the law of large numbers. Thus if $p > q$ then the walk tends to drift linearly towards $+\infty$, whilst if $p < q$ the drift is linear towards $-\infty$. If $p = q$ then $n^{-1}S_n \to 0$ in mean square and the walk remains 'centred' at its starting point 0.

Theorem 10B

The probability that a simple random walk ever revisits its starting point is given by

(6)
$$P(S_n = 0 \text{ for some } n = 1, 2, \ldots \mid S_0 = 0) = 1 - |p - q|.$$

Hence the walk is recurrent if and only if $p = q = \frac{1}{2}$.

Proof

We use generating functions in this proof. The basic step is as follows. We suppose that $S_0 = 0$ and we write

$$A_n = \{S_n = 0\}$$

for the event that the walk revisits its starting point at time n, and

$$B_n = \{S_n = 0, \quad S_k \neq 0 \text{ for } 1 \leq k \leq n-1\}$$

for the event that the first return of the walk to its starting point occurs at time n. If A_n occurs then exactly one of $B_1, B_2, \ldots, B_n$ occurs, giving by (1.9) that

(7)
$$P(A_n) = \sum_{k=1}^{n} P(A_n \cap B_k).$$

Now, $A_n \cap B_k$ is the event that the walk returns for the first time at time k and then returns again after a subsequent time $n-k$; hence

(8)
$$P(A_n \cap B_k) = P(B_k)P(A_{n-k}) \quad \text{for } 1 \leq k \leq n,$$

since transitions in disjoint intervals of time are independent of each other. We write $f_n = P(B_n)$ and $u_n = P(A_n)$, and substitute (8) into (7) to obtain

(9)
$$u_n = \sum_{k=1}^{n} f_k u_{n-k} \quad \text{for } n = 1, 2, \ldots.$$

In this equation, we know the u's from (4) and we want to find out about the f's. The form of the summation suggests the use of generating functions, and so we introduce the generating functions of the sequences of u's and f's,

$$U(s) = \sum_{n=0}^{\infty} u_n s^n, \qquad F(s) = \sum_{n=0}^{\infty} f_n s^n,$$

noting that $u_0 = 1$ and $f_0 = 0$. These sequences converge absolutely if $|s| < 1$, since $|u_n| \leq 1$ and $|f_n| \leq 1$ for each n. Now we multiply both sides of (9) by s^n and sum over n to find

(10)
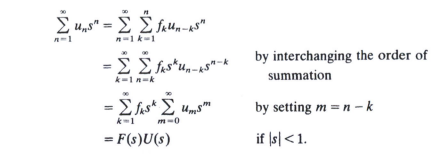

$$\sum_{n=1}^{\infty} u_n s^n = \sum_{n=1}^{\infty} \sum_{k=1}^{n} f_k u_{n-k} s^n$$

$$= \sum_{k=1}^{\infty} \sum_{n=k}^{\infty} f_k s^k u_{n-k} s^{n-k} \qquad \text{by interchanging the order of summation}$$

$$= \sum_{k=1}^{\infty} f_k s^k \sum_{m=0}^{\infty} u_m s^m \qquad \text{by setting } m = n-k$$

$$= F(s)U(s) \qquad \text{if } |s| < 1.$$

The left-hand side of this equation equals $U(s) - u_0 s^0 = U(s) - 1$, and so we have that

$$U(s) = U(s)F(s) + 1 \qquad \text{if } |s| < 1.$$

Hence

(11)
$$F(s) = 1 - U(s)^{-1} \qquad \text{if } |s| < 1.$$

Finally, by Theorem 10A,

$$U(s) = \sum_{n=0}^{\infty} u_n s^n$$

$$= \sum_{m=0}^{\infty} u_{2m} s^{2m} \qquad \text{since } u_n = 0 \text{ if } n \text{ is odd}$$

$$= \sum_{m=0}^{\infty} \binom{2m}{m} (pqs^2)^m$$

$$= (1 - 4pqs^2)^{-\frac{1}{2}} \qquad \text{if } |s| < 1,$$

by the negative binomial theorem (this last line is a calculation similar to that after (2.10)). This implies by (11) that

(12)
$$F(s) = 1 - (1 - 4pqs^2)^{\frac{1}{2}} \qquad \text{if } |s| < 1,$$

from which expression the f's may be found explicitly. To prove the theorem, we note that

$$P(S_n = 0 \text{ for some } n \geq 1 \mid S_0 = 0) = P(B_1 \cup B_2 \cup \cdots)$$

$$= f_1 + f_2 + \cdots \qquad \text{by (1.9)}$$

$$= \lim_{s \uparrow 1} \sum_{n=1}^{\infty} f_n s^n$$

$$= F(1) \qquad \text{by Abel's lemma†}$$

$$= 1 - (1 - 4pq)^{\frac{1}{2}} \qquad \text{by (12).}$$

Finally, remember that $p + q = 1$ to see that

$$(1 - 4pq)^{\frac{1}{2}} = [(p + q)^2 - 4pq]^{\frac{1}{2}}$$
$$= [(p - q)^2]^{\frac{1}{2}}$$
$$= |p - q|. \qquad \qquad \square$$

Thus, if $p = q = \frac{1}{2}$ then the walk is bound to return to its starting point. Let

$$T = \min\{n \geq 1 : S_n = 0\}$$

be the (random) time until the first return. We have shown that $P(T < \infty) = 1$ if $p = q = \frac{1}{2}$. Against this positive observation we must set the following negative one: although T is finite (with probability 1)

† For a statement of Abel's lemma, see the footnote in Section 4.3.

it has infinite mean in that $E(T) = \infty$. To see this, just note that

$$E(T) = \sum_{n=1}^{\infty} n f_n$$

$$= \lim_{s \uparrow 1} \sum_{n=1}^{\infty} n f_n s^{n-1} \qquad \text{by Abel's lemma}$$

$$= \lim_{s \uparrow 1} F'(s).$$

From (12), if $p = q = \frac{1}{2}$ then

$$F(s) = 1 - (1 - s^2)^{\frac{1}{2}} \qquad \text{if } |s| < 1$$

and

$$F'(s) = s(1 - s^2)^{-\frac{1}{2}}$$

$$\to \infty \quad \text{as } s \uparrow 1.$$

In other words, although a symmetric random walk is certain to return to its starting point, the expected value of the time which elapses before this happens is infinite.

Exercises 6. Consider a simple random walk with $p \neq q$. Show that, conditional on the walk returning to its starting point at some time, the expected number of steps taken before this occurs is $4pq \, [|p - q| \, (1 - |p - q|)]^{-1}$.
7. Show that a symmetric random walk revisits its starting point infinitely often with probability 1.
8. Show that a simple random walk starting from the origin is bound to visit the point 1 (with probability 1) if it is symmetric.

10.4 The Gambler's Ruin problem

The Gambler's Ruin problem is an old chestnut of probability textbooks. It concerns a game between two players, A and B say, who compete with each other as follows. Initially A possesses £a and B possesses £$(N - a)$, so that their total capital is £N. A coin is flipped repeatedly and comes up either heads with probability p or tails with probability q where $0 < p = 1 - q < 1$. Each time that heads turns up player B give £1 to player A, while each time tails turns up player A gives £1 to player B. This game continues until either A or B runs out of money. We record the state of play by noting down A's capital after each flip. Clearly the sequence of numbers which we write down is a random walk on the set $\{0, 1, \ldots, N\}$; this walk starts at the point a and follows a simple random walk until it reaches either 0 or N, at which time it stops; we speak of 0 and N as being *absorbing barriers* since the random walker sticks at whichever of these points it hits first. We shall say that A *wins the game* if the random walker is absorbed at N, and that B *wins the game* if the walker is absorbed at 0. It is fairly clear (see Exercise 9) that there is zero probability that the game will continue forever, so that either A or B (but not both) wins the game. What is the probability that A wins the game? The answer is given in the following theorem.

Theorem 10C   *Consider a simple random walk on $\{0, 1, \ldots, N\}$ with absorbing barriers at 0 and N. If the walk begins at the point a, where $0 \leq a \leq N$, then the probability $v(a)$ that the walk is absorbed at N is given by*

(13)
$$v(a) = \begin{cases} \dfrac{(q/p)^a - 1}{(q/p)^N - 1} & \text{if } p \neq q, \\ a/N & \text{if } p = q = \tfrac{1}{2}. \end{cases}$$

Thus, the probability that player A wins the game is given by equation (13). Our proof of this theorem uses the jargon of the Gambler's Ruin problem.

Proof   The first step of the argument is simple, and provides a difference equation for the numbers $v(0), v(1), \ldots, v(N)$. Let H be the event that the first flip of the coin shows heads. We use the partition theorem to see that

(14)
$$P(A \text{ wins}) = P(A \text{ wins} \mid H)P(H) + P(A \text{ wins} \mid H^c)P(H^c),$$

where, as usual, H^c is the complement of H. If H occurs then A's capital increases from £a to £$(a + 1)$, giving that $P(A \text{ wins} \mid H) = v(a + 1)$. Similarly $P(A \text{ wins} \mid H^c) = v(a - 1)$. We substitute these expressions into (14) to obtain

(15)
$$v(a) = pv(a + 1) + qv(a - 1) \qquad \text{for } 1 \leq a \leq N - 1.$$

This is a second-order difference equation subject to the boundary conditions

(16)
$$v(0) = 0, \qquad v(N) = 1,$$

since if A starts with £0 (or £N) then he has already lost (or won) the game. We solve (15) by the standard methods described in the appendix, obtaining as general solution

$$v(a) = \begin{cases} \alpha + \beta(q/p)^a & \text{if } p \neq q, \\ \alpha + \beta a & \text{if } p = q = \tfrac{1}{2}, \end{cases}$$

where α and β are constants which are found from the boundary conditions (16) as required. $\square$

There is another standard calculation which involves difference equations and arises from the Gambler's Ruin problem. This deals with the expected length of the game. Once again, we formulate this in terms of the related random walk.

Theorem 10D

Consider a simple random walk on $\{0, 1, \ldots, N\}$ with absorbing barriers at 0 and N. If the walk begins at the point a, where $0 \le a \le N$, then the expected number $e(a)$ of steps before the walk is absorbed at 0 or N is given by

(17)
$$e(a) = \begin{cases} \dfrac{1}{p-q}\left(N\dfrac{(q/p)^a - 1}{(q/p)^N - 1} - a\right) & \text{if } p \ne q, \\ a(N-a) & \text{if } p = q = \tfrac{1}{2}. \end{cases}$$

Thus, the expected number of flips of the coin before either A or B becomes bankrupt in the Gambler's Ruin problem is given by (17).

Proof

Let F be the number of flips of the coin before the game ends, and let H be the event that the first flip shows heads as before. By the partition theorem (Theorem 2C) we have that

(18)
$$E(F) = E(F \mid H)P(H) + E(F \mid H^c)P(H^c).$$

Now, if H occurs then, after the first flip of the coin, A's capital increases from £a to £$(a + 1)$, giving that $E(F \mid H) = 1 + e(a + 1)$ where the 1 counts the first flip and $e(a + 1)$ counts the mean number of subsequent flips. Similarly $E(F \mid H^c) = 1 + e(a - 1)$, and (18) becomes

$$e(a) = [1 + e(a + 1)]p + [1 + e(a - 1)]q$$

or

(19)
$$pe(a + 1) - e(a) + qe(a - 1) = -1 \quad \text{for } 1 \le a \le N - 1.$$

The boundary conditions for this second-order difference equation are

$$e(0) = e(N) = 0,$$

since the game is finished already if it starts in locations 0 or N. We solve (19) in the standard manner (as shown in the appendix) to obtain (17). □

Finally, what are A's fortunes if his opponent is infinitely rich? In practice this situation cannot arise, but the hypothetical situation may help us to understand the consequences of a visit to the casino at Monte Carlo. In this case, A can never defeat his opponent but he might at least hope to be spared ultimate bankruptcy so that he can play the game forever. The probability that he is ultimately bankrupted is given by our final theorem about random walks.

Theorem 10E *Consider a simple random walk on $\{0, 1, 2, \ldots\}$ with an absorbing barrier at 0. If the walk begins at the point a (≥ 0) then the probability $\pi(a)$ that the walk is ultimately absorbed at 0 is given by*

(20)
$$\pi(a) = \begin{cases} (q/p)^a & \text{if } p > q, \\ 1 & \text{if } p \leq q. \end{cases}$$

Thus, the probability that player A is able to play forever is strictly positive if and only if the odds are stacked in his favour at each flip of the coin. This justifies equation (3) in Section 10.1. An intuitive approach to this theorem is to think of this new game as the limit of the previous game as the total capital N tends to infinity while A's initial capital remains fixed at a. Thus,

$$P(A \text{ is bankrupted}) = \lim_{N \to \infty} [1 - v(a)]$$

where $v(a)$ is given by (13), and it is easy to see that the value of this limit is given by (20). There is a limiting argument here which requires some justification, but we shall use a different approach which avoids this.

Proof The sequence $\pi(0), \pi(1), \ldots$ satisfies the difference equation

(21)
$$\pi(a) = p\pi(a+1) + q\pi(a-1) \qquad \text{for } a = 1, 2, \ldots,$$

derived in exactly the same way as (15). The general solution is

$$\pi(a) = \begin{cases} \alpha + \beta(q/p)^a & \text{if } p \neq q, \\ \alpha + \beta a & \text{if } p = q = \frac{1}{2}, \end{cases}$$

where α and β are constants. Unfortunately we have only one boundary condition, which is that $\pi(0) = 1$. Using this condition we find that

$$\alpha + \beta = 1 \qquad \text{if } p \neq q,$$
$$\alpha = 1 \qquad \text{if } p = q,$$

and hence

(22)
$$\pi(a) = \begin{cases} \beta(q/p)^a + 1 - \beta & \text{if } p \neq q, \\ 1 + \beta a & \text{if } p = q = \frac{1}{2}, \end{cases}$$

for some $\beta \in \mathbb{R}$. Now, $\pi(a)$ is a probability and so $0 \leq \pi(a) \leq 1$ for all a. Hence, if $p = q$ then $\pi(a) = 1$ for all a. On the other hand, if $p < q$ then $(q/p)^a \to \infty$ as $a \to \infty$; thus, $\beta = 0$ if $p < q$, and we have proved that

(23) $$\pi(a) = 1 \quad \text{if } p \le q \qquad \text{for } a = 0, 1, 2, \ldots.$$

It is not quite so easy to find the correct value of β in (22) for the case when $p > q$, and we shall do this by calculating $\pi(1)$ explicitly for this case. For the remaining part of the proof we write $\pi(a) = \pi_{p,q}(a)$ in order to emphasize the roles of p and q; thus (23) reads

(24) $$\pi_{p,q}(a) = 1 \quad \text{if } p \le q \qquad \text{for } a = 0, 1, 2, \ldots.$$

Consider a simple random walk $T_0, T_1, \ldots$ starting from $T_0 = 0$ in which each step is to the right with probability p or to the left with probability q, and let C be the event that the walk revisits its starting point: $C = \{T_n = 0 \text{ for some } n \ge 1\}$. From Theorem 10B,

(25) $$P(C) = 1 - |p - q|.$$

On the other hand, the usual conditioning argument yields

$$P(C) = P(C \mid H)P(H) + P(C \mid H^c)P(H^c)$$

where H is the event that the first move is to the right. Now, $P(C \mid H) = \pi_{p,q}(1)$, since this is the probability that a walk ever reaches 0 having started from 1; also $P(C \mid H^c) = \pi_{q,p}(1)$, since this is the probability that a walk reaches 0 starting from -1, which is the probability that the 'mirror image' walk (in which each step of T is replaced by an opposite step) reaches 0 starting from 1. Thus

$$P(C) = p\pi_{p,q}(1) + q\pi_{q,p}(1),$$

which combines with (25) to give

(26) $$1 - |p - q| = p\pi_{p,q}(1) + q\pi_{q,p}(1).$$

If $p \ge q$ then $\pi_{q,p}(1) = 1$ by (24), and (26) becomes

$$1 - (p - q) = p\pi_{p,q}(1) + q,$$

implying that

$$\pi_{p,q}(1) = q/p \quad \text{if } p > q.$$

Substitute this into (22) to find that $\beta = 1$ if $p > q$, and the theorem is proved. $\qquad\square$

Exercises 9. Use Theorem 10C to show that, in the Gambler's Ruin problem, the probability that B wins the game is

$$\mu(a) = \begin{cases} \dfrac{(p/q)^{N-a} - 1}{(p/q)^N - 1} & \text{if } p \ne q, \\ (N - a)/N & \text{if } p = q = \frac{1}{2}, \end{cases}$$

where £$(N - a)$ is B's initial capital. Deduce that

$$P(A \text{ wins}) + P(B \text{ wins}) = 1,$$

giving that there is zero probability that the game will continue unresolved forever.

10.5 Problems

1. Two particles perform independent and simultaneous symmetric random walks starting from the origin. Show that the probability that they are at the same position after n steps is

$$\left(\frac{1}{2}\right)^{2n} \sum_{k=0}^{n} \binom{n}{k}^2.$$

Hence or otherwise show that

$$\sum_{k=0}^{n} \binom{n}{k}^2 = \binom{2n}{n}.$$

2. Consider a random walk on the integers with absorbing barriers at 0 and N in which, at each stage, the particle may jump one unit to the left (with probability α), remain where it is (with probability β), or jump one unit to the right (with probability γ), where α, β, $\gamma > 0$ and $\alpha + \beta + \gamma = 1$. If the particle starts from the point a where $0 \leq a \leq N$, show that the probability that it is absorbed at N is given by equation (13) with $p = 1 - q = \gamma/(\alpha + \gamma)$. Find the mean number of stages before the particle is absorbed at one or other of the barriers.

3. Two players play a gambling game as follows. Initially they have £N divided between them. At each stage Player A throws a fair die; if the outcome is an even number no money changes hands; if the outcome is a one, Player A gives Player B £1; finally if the outcome is a three or a five, Player A receives £1 from Player B. The game continues until one player has all £N. Let p_k be the probability that Player A wins (i.e. finally ends up with all £N) given that he starts the game with £k of the £N. Derive a difference equation relating p_k, p_{k-1} and p_{k+1} for $k = 1, 2, \ldots, N - 1$. Determine p_k by solving this equation subject to suitable boundary conditions. Let t_k be the expected total number of throws in the game given that Player A starts with £k. Write down a difference equation relating t_{k-1}, t_k, and t_{k+1} for $k = 1, 2, \ldots, N - 1$.
(Bristol 1983)

4. A particle performs a random walk on the integers, $-N$, $-N + 1, \ldots, N - 1$, N, in which it is absorbed if it reaches $-N$ or N, where $N > 1$. The probability of a step of size $+1$ is p, and the probability of a step of size -1 is $q = 1 - p$, with $0 < p < 1$. Suppose that the particle starts at 0. By conditioning on the first step and using Theorem 10C, or otherwise, show that when $p \neq q$ the probability of the particle being absorbed at N or $-N$ before returning to 0 is $(p - q)(p^N + q^N)/(p^N - q^N)$. What is this probability when $p = q$? (Oxford 1983M)

5. Consider a random walk on the integers in which the particle moves

either two units to the right (with probability p) or one unit to the left (with probability $q = 1 - p$) at each stage, where $0 < p < 1$. There is an absorbing barrier at 0 and the particle starts at the point a (> 0). Show that the probability $\pi(a)$ that the particle is ultimately absorbed at 0 satisfies the difference equation

$$\pi(a) = p\pi(a + 2) + q\pi(a - 1) \qquad \text{for } a = 1, 2, \ldots,$$

and deduce that $\pi(a) = 1$ if $p \le \frac{1}{3}$. Suppose that the particle is absorbed whenever it hits either N or $N + 1$. Find the probability $\pi_N(a)$ that it is absorbed at 0 rather than at N or $N + 1$, having started at a where $0 \le a \le N + 1$; deduce that, as $N \to \infty$,

$$\pi_N(a) \to \begin{cases} 1 & \text{if } p \le \frac{1}{3}, \\ \theta^a & \text{if } p > \frac{1}{3}, \end{cases}$$

where $\theta = \frac{1}{2}\{\sqrt{[1 + (4q/p)]} - 1\}$.

6. Consider a simple random walk with an absorbing barrier at 0 and a 'retaining' barrier at N; that is to say, the walk is not allowed to pass to the right of N, so that its position S_n at time n satisfies

$$P(S_{n+1} = N \mid S_n = N) = p, \qquad P(S_{n+1} = N - 1 \mid S_n = N) = q.$$

Set up a difference equation for the mean number $e(a)$ of jumps of the walk until absorption at 0, starting from a where $0 \le a \le N$. Deduce that

$$e(a) = a(2N - a + 1) \quad \text{if } p = q = \frac{1}{2},$$

and find $e(a)$ if $p \ne q$.

7. Let N be the number of times that an asymmetric simple random walk revisits its starting point. Show that N is a discrete random variable with mass function

$$P(N = k) = \alpha(1 - \alpha)^k \qquad \text{for } k = 0, 1, 2, \ldots,$$

where $\alpha = |p - q|$ and p is the probability that each step of the walk is to the right.

8. Consider the two-dimensional random walk of Exercise 5 at the end of Section 10.2, in which a particle inhabits the integer points $\{(i, j) : i, j = \ldots, -1, 0, 1, \ldots\}$ of the plane, moving rightwards, upwards, leftwards, or downwards with respective probabilities p, q, r, and s at each step, where $p, q, r, s > 0$ and $p + q + r + s = 1$. Let S_n be the particle's position after n steps, and suppose that $S_0 = (0, 0)$. Let v_n be the probability that $S_n = (0, 0)$, and prove that

$$v_{2m} = \binom{2m}{m}^2 (\tfrac{1}{4})^{2m} \qquad \text{if } p = q = r = s = \tfrac{1}{4}.$$

Use Stirling's formula to show that

$$\sum_{n=0}^{\infty} v_n = \infty \qquad \text{if } p = q = r = s = \tfrac{1}{4}.$$

Deduce directly that the symmetric random walk in two dimensions is recurrent.

9. Here is another way of approaching the symmetric random walk in two dimensions of Problem 8. Make the following change of variables. If $S_n = (i, j)$, set $X_n = i + j$ and $Y_n = i - j$; this is equivalent to rotating the axes through an angle of $\frac{1}{4}\pi$. Show that $X_0, X_1, \ldots$ and $Y_0, Y_1, \ldots$ are independent symmetric random walks in one dimension. Deduce from Theorem 10A that

$$P(S_{2m} = 0 \mid S_0 = 0) = \left[\binom{2m}{m}\left(\frac{1}{2}\right)^{2m}\right]^2,$$

where $\mathbf{0} = (0, 0)$.

10. In the two-dimensional random walk of Problem 8, let D_n be the euclidean distance between the origin and S_n. Prove that if the walk is symmetric

$$E(D_n^2) = E(D_{n-1}^2) + 1 \qquad \text{for } n = 1, 2, \ldots,$$

and deduce that $E(D_n^2) = n$.

*11. A particle performs a random walk on the integers starting at the origin. At discrete intervals of time it takes a step of unit size. The steps are independent and equally likely to be in the positive or negative direction. Determine the probability generating function of the time at which the particle first reaches the integer n.

   In a two-dimensional random walk a particle can be at any of the points (x, y) which have integer coordinates. The particle starts at $(0, 0)$ and at discrete intervals of time takes a step of unit size. The steps are independent and equally likely to be to any of the four nearest points. Show that the probability generating function of the time taken to reach the line $x + y = m$ is $\{[1 - \sqrt{(1 - s^2)}]/s\}^m$, $|s| \le 1$. Let (X, Y) be the random point on the line $x + y = m$ which is reached first of all. What is the probability generating function of $X - Y$?   (Oxford 1979F)

12. Consider a symmetric random walk on the integer points of the cubic lattice $\{(i, j, k) : i, j, k = \ldots, -1, 0, 1, \ldots\}$ in three dimensions, in which the particle moves to one of its six neighbouring positions, chosen with equal probability $\frac{1}{6}$, at each stage. Show that the probability w_n that the particle revisits its starting point at the nth stage is given by

$$w_n = 0 \qquad \text{if } n \text{ is odd},$$

$$w_{2m} = \left(\frac{1}{2}\right)^{2m}\binom{2m}{m} \sum_{\substack{(i,j,k): \\ i+j+k=m}} \left(\frac{m!}{3^m i! \, j! \, k!}\right)^2 \qquad \text{if } n = 2m \text{ is even}.$$

Use Stirling's formula to show that

$$\sum_{n=0}^{\infty} w_n < \infty.$$

Deduce that the symmetric random walk in three dimensions is transient (the general argument in Problem 7 may be useful here again).

13. Show that the symmetric random walk on the integer points $\{(i_1, i_2, \ldots, i_d) : i_j = \ldots, -1, 0, 1, \ldots, j = 1, 2, \ldots, d\}$ is recurrent if $d = 1, 2$ and transient if $d \ge 3$. You should use the results of Problems 8 and 12, and you need do no more calculations.

14. The generating-function argument used to prove Theorem 10B has a powerful application to the general theory of 'recurrent events'. Let η be an event which may or may not happen at each of the time points $0, 1, 2, \ldots$ (η may be the visit of a random walk to its starting point, or a visit to the dentist, or a car accident outside the department); we shall suppose that η occurs at time 0. Suppose further that the intervals between successive occurrences of η are independent identically distributed random variables $X_1, X_2, \ldots$ each having mass function

$$P(X = k) = f_k \qquad \text{for } k = 1, 2, \ldots,$$

so that η occurs at the times $0, X_1, X_1 + X_2, X_1 + X_2 + X_3, \ldots$. There may exist a time after which η never occurs; that is to say, there may be an X_i which takes the value ∞, and we allow for this by requiring only that $f = f_1 + f_2 + \cdots$ satisfies $f \leq 1$, and we set

$$P(X = \infty) = 1 - f.$$

We call η *recurrent* if $f = 1$ and *transient* if $f < 1$. Let u_n be the probability that η occurs at time n. Show that the generating functions

$$F(s) = \sum_{n=1}^{\infty} f_n s^n, \qquad U(s) = \sum_{n=0}^{\infty} u_n s^n$$

are related by

$$U(s) = U(s)F(s) + 1,$$

and deduce that η is recurrent if and only if $\sum_n u_n = \infty$.

15. The university buys lightbulbs which have random lifetimes. If the bulb in my office fails on day n then it is replaced by a new bulb which lasts for a random number of days, after which it is changed, and so on. We assume that the lifetimes of the bulbs are independent random variables $X_1, X_2, \ldots$ each having mass function

$$P(X = k) = (1 - \alpha)\alpha^{k-1} \qquad \text{for } k = 1, 2, \ldots,$$

where α satisfies $0 < \alpha < 1$. A new lightbulb is inserted in its socket on day 0. Show that the probability that the bulb has to be changed on day n is $1 - \alpha$, independently of n.

11
Random processes in continuous time

11.1 Life at a telephone switchboard

Branching processes and random walks are two examples of random processes; each is a random sequence, and we call them discrete-time processes since they involve repeated observations at the times $n = 0$, $1, 2, \ldots$. Many other processes involve observations which are made continuously as time passes, and such processes are called *continuous-time* processes. Rather than being a family $(Z_n : n = 0, 1, 2, \ldots)$ of random variables indexed by the non-negative integers, a continuous-time process is a family $Z = (Z_t : t \geq 0)$ of random variables indexed by the larger set $[0, \infty)$ where we think of Z_t as being the value of the process at time t. The general theory of continuous-time processes is rather deep and quite difficult, but most of the main difficulties are avoided if we restrict our attention to processes which take integer values only, that is, processes for which Z_t is a (random) integer for each t, and all our examples are of this form. The principal difference between studying such continuous-time processes and studying discrete-time processes is merely that which arises in moving from the integers to the reals; instead of establishing *recurrence* equations and *difference* equations (viz. (9.5) and (10.15)), we shall establish *differential* equations.

Here is our basic example. Bill is the head porter at the Grand Hotel, and part of his job is to answer incoming telephone calls. He cannot predict with certainty when the telephone will ring, in fact from his point of view calls seem to arrive at random. We make two simplifying assumptions about these calls. First, we assume that Bill deals with every call instantaneously, so that no call is lost unless it arrives at exactly the same moment as another (in practice Bill has to get to the telephone and speak, and this takes time—more complicated models will take account of this). Secondly, we assume that calls arrive 'homogeneously' in time in the sense that the chance that the telephone rings during any given period of time depends only upon the length of this period (this is an absurd assumption of course, but it may be valid for certain portions of the day). We describe *time* by a parameter t taking values in $[0, \infty)$, and propose the following model for the arrivals of telephone calls at the switchboard. We let N_t represent the number of calls which have arrived in the time-interval $[0, t]$: that is, N_t is the number of incoming calls which Bill has handled up to and including time t. We suppose that the random process $N = (N_t : t \geq 0)$ evolves in such a way that the following conditions are valid:

(1) N_t is a random variable taking values in $\{0, 1, 2, \ldots\}$,

(2) $N_0 = 0$,

(3) $N_s \leq N_t$ if $s \leq t$,

(4)   *independence*: if $0 \leq s < t$ then the number of calls which arrive during the time-interval $(s, t]$ is independent of the arrivals of calls prior to time s,

(5)   *arrival rate*: there exists a number λ (>0), called the *arrival rate*, such that†, for small positive h,
         (i) $P(N_{t+h} = n + 1 \mid N_t = n) = \lambda h + o(h)$,
         (ii) $P(N_{t+h} = n \mid N_t = n) = 1 - \lambda h + o(h)$.
         Condition (5) deserves a discussion. It postulates that the probability that a call arrives in some short interval $(t, t + h]$ is approximately a linear function of h, and that this approximation becomes better and better as h becomes smaller and smaller. It follows from (5i, ii) that the chance of two or more calls in the interval $(t, t + h]$ satisfies

$$P(N_{t+h} \geq n + 2 \mid N_t = n) = 1 - P(N_{t+h} \text{ equals } n \text{ or } n + 1 \mid N_t = n)$$
$$= 1 - [\lambda h + o(h)] - [1 - \lambda h + o(h)]$$
$$= o(h),$$

so that the only two possible events having significant probabilities (that is, with probability greater than $o(h)$) involve either *no* call arriving in $(t, t + h]$ or exactly *one* call arriving in this time-interval.

   This is our model for the arrival of telephone calls. It is a primitive model based on the idea of random arrivals, and obtained with the aid of various simplifying assumptions. For a reason which will soon be clear, the random process $N = (N_t : t \geq 0)$ which we have con-structed is called a *Poisson process with rate* λ; Poisson processes may be used to model many phenomena, such as
(a) the arrival of customers in a shop,
(b) the clicks emitted by a Geiger counter as it records the detection of radioactive particles,

---

† Remember that $o(h)$ denotes some function of h which is of smaller order of magnitude than h as $h \to 0$. More precisely, we write $f(h) = o(h)$ if $h^{-1}f(h) \to 0$ as $h \to 0$. The term $o(h)$ generally represents different functions of h at each appearance. Thus, for example, $o(h) + o(h) = o(h)$.

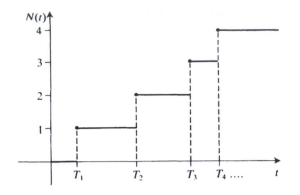

Fig. 11.1 A sketch of the graph of a Poisson process

(c) the incidence of deaths in a small town with a reasonably stable population (neglecting seasonal variations).

It provides an exceptionally good model for the emission of radioactive particles when the source has a long half-life and is decaying slowly.

We may represent the outcomes of a Poisson process N by a graph of N_t against t (see Fig. 11.1). Let T_i be the time at which the ith call arrives; that is

(6)
$$T_i = \inf\{t : N_t = i\}.$$

Then $T_0 = 0$, $T_0 \leq T_1 \leq T_2 \leq \cdots$ and $N_t = i$ if t lies in the interval $[T_i, T_{i+1})$. We note that $T_0, T_1, T_2, \ldots$ is a sequence of random variables whose values determine the process N completely: if we know the T's, then N_t is given by

$$N_t = \max\{n : T_n \leq t\}.$$

The sequence of T's may be thought of as the 'inverse process' of N.

Conditions (1)–(5) are our postulates for a Poisson process N; in the next two sections we present some consequences of these postulates, answering such questions as
(a) what is the mass function of N_t, for a given value of t?
(b) what can be said about the distribution of the sequence T_0, $T_1, \ldots$ of times at which calls arrive?

Exercises   1. If N is a Poisson process with rate λ, show that
$$P(N_{t+h} = 0) = [1 - \lambda h + o(h)]P(N_t = 0)$$
for small positive values of h. Hence show that $p(t) = P(N_t = 0)$ satisfies

the differential equation

$$p'(t) = -\lambda p(t).$$

Solve this equation to find $p(t)$.

2. Suppose that telephone calls arrive at the exchange in the manner of a Poisson process $N = (N_t : t \geq 0)$ with rate λ, and suppose that the equipment is faulty so that each incoming call fails to be recorded with probability q (independently of all other calls). If N_t' is the number of calls recorded by time t, show that $N' = (N_t' : t \geq 0)$ is a Poisson process with rate $\lambda(1 - q)$.

3. Two independent streams of telephone calls arrive at the exchange, the first being a Poisson process with rate λ and the second being a Poisson process with rate μ. Show that the combined stream of calls is a Poisson process with rate $\lambda + \mu$.

11.2   Poisson processes

A Poisson process with rate λ is a random process which satisfies postulates (1)–(5). Our first result explains this term.

Theorem 11A   *For each $t > 0$, the random variable N_t has the Poisson distribution with parameter λt. That is, for all $t > 0$,*

(7)
$$P(N_t = k) = \frac{1}{k!}(\lambda t)^k e^{-\lambda t} \qquad for\ k = 0, 1, 2, \ldots.$$

It follows from (7) that the mean and variance of $N(t)$ grow linearly in t as t increases; that is,

(8)
$$E(N_t) = \lambda t, \quad \mathrm{var}(N_t) = \lambda t, \qquad for\ t > 0.$$

Proof   Just as we set up difference equations for discrete-time processes, here we set up 'differential-difference' equations. Let

$$p_k(t) = P(N_t = k).$$

Fix $t \geq 0$ and let h be small and positive. The basic step is to express N_{t+h} in terms of N_t as follows. We use the partition theorem (Theorem 1B) to see that, if $k \geq 1$,

(9)
$$P(N_{t+h} = k) = \sum_{i=0}^{k} P(N_{t+h} = k \mid N_t = i)P(N_t = i)$$

$$= P(N_{t+h} = k \mid N_t = k - 1)P(N_t = k - 1)$$
$$+ P(N_{t+h} = k \mid N_t = k)P(N_t = k) + o(h) \qquad \text{by (5)}$$
$$= [\lambda h + o(h)]P(N_t = k - 1) + [1 - \lambda h + o(h)]P(N(t) = k)$$
$$+ o(h) \quad \text{by (5)}$$
$$= \lambda h P(N_t = k - 1) + (1 - \lambda h)P(N_t = k) + o(h)$$

giving that

(10)
$$p_k(t+h) - p_k(t) = \lambda h[p_{k-1}(t) - p_k(t)] + o(h),$$

valid for $k = 1, 2, \ldots$. We divide both sides of (10) by h and take the limit as $h \downarrow 0$ to obtain

(11)
$$p'_k(t) = \lambda p_{k-1}(t) - \lambda p_k(t) \qquad \text{for } k = 1, 2, \ldots,$$

where $p'_k(t)$ is the derivative of $p_k(t)$ with respect to t. If $k = 0$, then (9) becomes

$$P(N_{t+h} = 0) = P(N_{t+h} = 0 \mid N_t = 0)P(N_t = 0)$$
$$= (1 - \lambda h)P(N_t = 0) + o(h),$$

giving in the same way

(12)
$$p'_0(t) = -\lambda p_0(t).$$

Equations (11) and (12) are a system of 'differential-difference' equations for the functions $p_0(t), p_1(t), \ldots$, and we wish to solve them subject to the boundary condition $N_0 = 0$, which is equivalent to the condition

(13)
$$p_k(0) = \begin{cases} 1 & \text{if } k = 0, \\ 0 & \text{if } k \neq 0. \end{cases}$$

We present two ways of solving this family of equations.

Solution A *By induction.* Equation (12) involves $p_0(t)$ alone. Its general solution is $p_0(t) = Ae^{-\lambda t}$ and the arbitrary constant A is found from (13) to equal 1; hence

(14)
$$p_0(t) = e^{-\lambda t} \qquad \text{for } t \geq 0.$$

Substitute this into (11) with $n = 1$ to obtain

$$p'_1(t) + \lambda p_1(t) = \lambda e^{-\lambda t}$$

which, with the aid of an integrating factor and the boundary condition, yields

(15)
$$p_1(t) = \lambda t e^{-\lambda t} \qquad \text{for } t \geq 0.$$

Continue in this way to find that

$$p_2(t) = \tfrac{1}{2}(\lambda t)^2 e^{-\lambda t}.$$

Now guess the general solution (7) and prove it from (11) by induction.

Solution B

By generating functions. This method is nicer and has many more applications than the above solution by induction. We use the probability generating function of N_t, namely

$$G(s, t) = \mathsf{E}(s^{N_t}) = \sum_{k=0}^{\infty} p_k(t)s^k.$$

We multiply both sides of (11) by s^k and sum over the values $k = 1, 2, \ldots$ to find that

$$\sum_{k=1}^{\infty} p_k'(t)s^k = \lambda \sum_{k=1}^{\infty} p_{k-1}(t)s^k - \lambda \sum_{k=1}^{\infty} p_k(t)s^k.$$

Add (12) to this in the obvious way and note that

$$\sum_{k=1}^{\infty} p_{k-1}(t)s^k = sG(s, t)$$

and (plus or minus a dash of mathematical rigour)

$$\sum_{k=0}^{\infty} p_k'(t)s^k = \frac{\partial G}{\partial t},$$

to obtain

(16)
$$\frac{\partial G}{\partial t} = \lambda sG - \lambda G,$$

a differential equation with the boundary condition

(17)
$$G(s, 0) = \sum_{k=0}^{\infty} p_k(0)s^k = 1 \qquad \text{by (13).}$$

Equation (16) may be written in the form

$$\frac{1}{G}\frac{\partial G}{\partial t} = \lambda(s - 1);$$

this looks like a partial differential equation, but for each given value of s it may be integrated in the usual manner with respect to t, giving that

$$\log G = \lambda t(s - 1) + A(s)$$

where $A(s)$ is an arbitrary function of s. Use (17) to find that $A(s) = 0$ for all s, and hence

$$G(s, t) = e^{\lambda t(s-1)} = \sum_{k=0}^{\infty} \left(\frac{1}{k!}(\lambda t)^k e^{-\lambda t}\right)s^k.$$

Reading off the coefficient of s^k, we have that

$$p_k(t) = \frac{1}{k!}(\lambda t)^k e^{-\lambda t}$$

as required. □

Exercises 4. If N is a Poisson process with rate λ, show that var$(N_t/t) \to 0$ as $t \to \infty$.

5. If N is a Poisson process with rate λ, show that, for $t > 0$,

$$P(N_t \text{ is even}) = e^{-\lambda t} \cosh \lambda t,$$
$$P(N_t \text{ is odd}) = e^{-\lambda t} \sinh \lambda t.$$

6. If N is a Poisson process with rate λ, show that the moment generating function of

$$U_t = \frac{N_t - E(N_t)}{\sqrt{\text{var}(N_t)}}$$

is

$$M_t(x) = E(e^{xU_t})$$
$$= \exp[-x\sqrt{(\lambda t)} + \lambda t(e^{x/\sqrt{(\lambda t)}} - 1)].$$

Deduce that, as $t \to \infty$,

$$P(U_t \le u) \to \int_{-\infty}^{u} \frac{1}{\sqrt{(2\pi)}} e^{-\frac{1}{2}v^2} dv \qquad \text{for } u \in \mathbb{R};$$

this is the central limit theorem for a Poisson process.

11.3 Inter-arrival times and the exponential distribution

Let N be a Poisson process with rate λ as before. The *arrival times* $T_0, T_1, \ldots$ of N are defined as before by $T_0 = 0$ and

(18)
$$T_i = \inf\{t : N_t = i\} \qquad \text{for } i = 1, 2, \ldots;$$

in other words, T_i is the time of arrival of the ith telephone call. The *inter-arrival times* $X_1, X_2, \ldots$ are the times between successive arrivals; that is,

(19)
$$X_i = T_i - T_{i-1} \qquad \text{for } i = 1, 2, \ldots.$$

The distributions of the X's are very simple to describe.

Theorem 11B *In a Poisson process with rate λ, the inter-arrival times $X_1, X_2, \ldots$ are independent random variables, each having the exponential distribution with parameter λ.*

This result demonstrates an intimate link between the postulates (1)–(5) for a Poisson process and the exponential distribution.

Theorem 11B is only the tip of the iceberg; a deeper investigation into continuous-time random processes would reveal that the exponential distribution is a cornerstone for processes which satisfy an independence condition such as (4). The reason for this is that the exponential distribution is the only continuous distribution which has the so-called *lack-of-memory property*.

(20)   *Lack-of-memory property.* A positive random variable X is said to have this property if

(21) $P(X > u + v \mid X > u) = P(X > v)$ for all $u, v \geq 0$.

Thinking about X as the time which elapses before some event, A say, occurs, then condition (21) requires that if A has not occurred by time u then the time which elapses subsequently (between u and the occurrence of A) does not depend on the value of u: 'the random variable X does not remember how old it is when it plans its future'.

Theorem 11C   *The random variable X has the lack-of-memory property if and only if X is exponentially distributed.*

Proof   If X is exponentially distributed with parameter λ then, if $u, v \geq 0$,

$$P(X > u + v \mid X > u) = \frac{P(X > u + v \text{ and } X > u)}{P(X > u)}$$

$$= \frac{P(X > u + v)}{P(X > u)} \qquad \text{since } u \leq u + v$$

$$= \frac{e^{-\lambda(u+v)}}{e^{-\lambda u}} \qquad \text{by (5.15)}$$

$$= e^{-\lambda v}$$

$$= P(X > v),$$

so that X has the lack-of-memory property. Conversely, suppose that X has the lack-of-memory property and define $G(u) = P(X > u)$ for $u \geq 0$. The left-hand side of (21) is

$$P(X > u + v \mid X > u) = \frac{P(X > u + v)}{P(X > u)} \qquad \text{as before}$$

$$= \frac{G(u + v)}{G(u)},$$

and so G satisfies the 'functional equation'

(22) $G(u + v) = G(u)G(v)$ for $u, v \geq 0$.

$G(u)$ is a non-increasing function of u, and all non-zero non-increasing solutions of (22) are of the form

(23)
$$G(u) = e^{-\lambda u} \qquad \text{for } u \geq 0$$

where λ is some constant. It is an interesting exercise in analysis to derive (23) from (22), and we suggest that the reader does this for himself. First, use (22) to show that $G(n) = [G(1)]^n$ for $n = 0$, 1, 2, ..., and then deduce that $G(u) = [G(1)]^u$ for all non-negative rationals u, and finally use monotonicity to extend this from the rationals to the reals. □

Sketch proof of Theorem 11B

We sketch this only. Consider X_1 first. Clearly

$$P(X_1 > u) = P(N_u = 0) \qquad \text{for } u \geq 0,$$

and Theorem 11A gives

$$P(X_1 > u) = e^{-\lambda u} \qquad \text{for } u \geq 0,$$

so that X_1 has the exponential distribution with parameter λ. From the independence assumption (4), arrivals in the interval $(0, X_1]$ are independent of arrivals subsequent to X_1, and it follows that the 'waiting time' X_2 for the next arrival after X_1 is independent of X_1. Furthermore, arrivals occur 'homogeneously' in time (since the probability of an arrival in $(t, t + h]$ does not depend on t but only on h—remember (5)), giving that X_2 has the same distribution as X_1. Similarly, all the X's are independent with the same distribution as X_1. □

The argument of the proof above is incomplete, since the step involving independence deals with an interval $(0, X_1]$ of *random* length. It is not entirely a trivial task to make this step rigorous, and it is for this reason that the proof is only sketched.

We have shown above that if N is a Poisson process with parameter λ then the times X_1, X_2, ... between arrivals in this process are independent and identically distributed with the exponential distribution, parameter λ. This conclusion characterizes the Poisson process, in the sense that Poisson processes are the only 'arrival processes' with this property. More properly, we have the following. Let X_1^*, X_2^*, ... be independent random variables each having the exponential distribution with parameter λ (>0), and suppose that the telephone at the Grand Hotel is replaced by a very special new model which is programmed to ring at the times

$$T_1^* = X_1^*, \; T_2^* = X_1^* + X_2^*, \; T_3^* = X_1^* + X_2^* + X_3^*, \ldots,$$

so that the time which elapses between the $(i - 1)$th and the ith call equals X_i^*. Let

$$N_t^* = \max\{k : T_k^* \leq t\}$$

be the number of calls which have arrived by time t. Then the process $N^* = (N_t^* : t \geq 0)$ is a Poisson process with rate λ, so that from Bill's point of view the new telephone rings in exactly the same way (statistically speaking) as the old model.

Example 24   Suppose that buses for the city centre arrive at the bus-stop on the corner in the manner of a Poisson process. Knowing this, David expects to wait an exponentially distributed period of time before a bus will pick him up. If he arrives at the bus-stop and Doris tells him that she has been waiting 50 minutes already, then this is neither good nor bad news for him since the exponential distribution has the lack-of-memory property. Similarly, if he arrives just in time to see a bus departing, then he need not worry that his wait will be longer than usual. These properties are characteristics of the Poisson process.   □

Exercises   7. Let M and N be independent Poisson processes, M having rate λ and N having rate μ. Use the result of Problem 4 in Section 6.9 to show that the process $M + N = (M_t + N_t : t \geq 0)$ is a Poisson process with rate $\lambda + \mu$. Compare this method with the method of Exercise 3.

8. If T_i is the time of the ith arrival in the Poisson process N, show that $N_t < k$ if and only if $T_k > t$. Use Theorem 11B and the central limit theorem (Theorem 8E) to deduce that, as $t \to \infty$,

$$P\left(\frac{N_t - \lambda t}{\sqrt{(\lambda t)}} \leq u\right) \to \int_{-\infty}^{u} \frac{1}{\sqrt{(2\pi)}} e^{-\frac{1}{2}v^2} \, dv \qquad \text{for } u \in \mathbb{R}.$$

9. Calls arrive at a telephone exchange in the manner of a Poisson process with rate λ, but the operator is working to rule and answers every other call only. What is the common distribution of the time-intervals which elapse between successive calls which elicit responses.

11.4   Population growth, and the simple birth process

The ideas and methods of the last sections have many applications, one of which is a continuous-time model for population growth. We are thinking here of a simple model for phenomena such as the progressive cell-division of an amoeba, and we shall formulate the process in these terms. A hypothetical type of amoeba multiplies in the following way. At time $t = 0$ there are a number (I, say) of initial amoebas in a large pond and, as time passes, these amoebas multiply in number by the process of progressive cell-division; when an amoeba divides, the single parent-amoeba is replaced by exactly two identical copies of itself. So the number of amoebas in the pond grows as time passes, but we cannot say with certainty how many there will be in the future since cell-divisions occur irregularly. We assume here that cell-divisions occur at *random* times (rather as telephone calls arrived at random in the earlier model) and make the following two assumptions about this process:

(25)† *division rate*: each amoeba present in the pond at time t has a chance of dividing during the short time-interval $(t, t+h)$; there exists a constant λ (>0), called the *birth rate*, such that the probability that any such amoeba
(i) divides *once* in the time-interval $(t, t+h]$ equals $\lambda h + o(h)$,
(ii) does *not* divide in the time-interval $(t, t+h]$ equals $1 - \lambda h + o(h)$,

(26)† *independence*: for each amoeba at time t, all the future divisions of this amoeba occur independently both of its past history and of the activities (past and future) of all other amoebas present at time t.

From (25i, ii), each amoeba present at time t has probability $\lambda h + o(h)$ of giving rise to two amoebas by time $t + h$, probability $1 - \lambda h + o(h)$ of giving rise to one amoeba (itself) by time $t + h$, and consequently probability $o(h)$ of giving rise to more than two amoebas by $t + h$. Let M_t be the number of amoebas present at time t. From the previous observations, it is not difficult to write down the way in which the distribution of M_{t+h} depends on M_t. Suppose that $M_t = k$. Then $M_{t+h} \geq k$ and

(27)
$$P(M_{t+h} = k \mid M_t = k) = P(\text{no division})$$
$$= [1 - \lambda h + o(h)]^k \qquad \text{by (25) and (26)}$$
$$= 1 - \lambda k h + o(h).$$

Also

(28)
$$P(M_{t+h} = k + 1 \mid M_t = k) = P(\text{exactly one division})$$
$$= \binom{k}{1}[\lambda h + o(h)][1 - \lambda h + o(h)]^{k-1}$$
$$= \lambda k h + o(h),$$

† In practice, amoebas and bacteria multiply at rates which depend upon their environments. Although there is considerable variation between the life-cycles of cells in the same environment, these generally lack the degrees of homogeneity and independence which we postulate here.

since there are k possible choices for the cell-division. Finally,

(29)
$$P(M_{t+h} \geq k + 2 \mid M_t = k) = 1 - P(M_{t+h} \text{ is } k \text{ or } k + 1 \mid M_t = k)$$
$$= 1 - [\lambda kh + o(h)] - [1 - \lambda kh + o(h)]$$
$$= o(h).$$

Consequently the process M evolves in very much the same general way as the Poisson process N, in that if $M_t = k$ then either $M_{t+h} = k$ or $M_{t+h} = k + 1$ with probability $1 - o(h)$; the big difference between M and N lies in a comparison of (28) and (5i): the 'rate' at which M increases is proportional to M itself, whereas a Poisson process increases at a constant 'rate'. The process M is called a *simple (linear) birth* process or a *pure birth* process. We treat it with the same techniques which we used for the Poisson process.

Theorem 11D

If $M_0 = I$ and $t > 0$ then

(30)
$$P(M_t = k) = \binom{k-1}{I-1} e^{-I\lambda t}(1 - e^{-\lambda t})^{k-I} \qquad \text{for } k = I, I+1, \ldots .$$

Proof

Let
$$p_k(t) = P(M_t = k),$$

as before. We establish differential–difference equations for $p_I(t), p_{I+1}(t), \ldots$ in just the same way as we found (11) and (12). Thus we have from the partition theorem that, for $h > 0$,

$$P(M_{t+h} = k) = \sum_i P(M_{t+h} = k \mid M_t = i)P(M_t = i)$$
$$= [1 - \lambda kh + o(h)]P(M_t = k)$$
$$\qquad + [\lambda(k - 1)h + o(h)]P(M_t = k - 1) + o(h)$$

by (27)–(29), giving that

$$p_k(t + h) - p_k(t) = \lambda(k - 1)hp_{k-1}(t) - \lambda khp_k(t) + o(h).$$

Divide this equation by h and take the limit as $h \to 0$ to obtain

(31)
$$p_k'(t) = \lambda(k - 1)p_{k-1}(t) - \lambda kp_k(t) \qquad \text{for } k = I, I+1, \ldots ;$$

the equation for $p_I'(t)$ involves $p_{I-1}(t)$, and we note that $p_{I-1}(t) = 0$ for all t. We can solve (31) recursively subject to the boundary condition

(32)
$$p_k(0) = \begin{cases} 1 & \text{if } k = I, \\ 0 & \text{if } k \neq I, \end{cases}$$

(that is to say, first find $p_I(t)$, then $p_{I+1}(t)$, ...) and then (30) follows by induction.

We note that the method of generating functions works also. If we multiply through (31) by s^k and sum over k, we obtain the partial differential equation

(33)
$$\frac{\partial G}{\partial t} = \lambda s(s-1)\frac{\partial G}{\partial s}$$

where $G = G(s, t)$ is the generating function

$$G(s, t) = \sum_{k=1}^{\infty} p_k(t)s^k.$$

It is not difficult to solve this differential equation subject to the boundary condition $G(s, 0) = s^I$, but we do not require such skills from the reader. □

The mean and variance of M_t may be calculated directly from (30) in the usual way. These calculations are a little complicated since M_t has the negative binomial distribution, and it is simpler to use the following trick. Writing

$$\mu(t) = \mathsf{E}(M_t) = \sum_{k=I}^{\infty} kp_k(t)$$

we have, by differentiating blithely through the summation sign, that

(34)
$$\mu'(t) = \sum_{k=I}^{\infty} kp'_k(t)$$

$$= \sum_{k=I}^{\infty} k[\lambda(k-1)p_{k-1}(t) - \lambda kp_k(t)]$$

from (31). We collect the coefficients of $p_k(t)$ together here, to see that

(35)
$$\mu'(t) = \lambda \sum_{k=I}^{\infty} [(k+1)kp_k(t) - k^2p_k(t)]$$

$$= \lambda \sum_{k=I}^{\infty} kp_k(t)$$

$$= \lambda\mu(t),$$

which is a differential equation in μ with the boundary condition

$$\mu(0) = \mathsf{E}(M_0) = I.$$

This differential equation has solution

(36)
$$\mu(t) = Ie^{\lambda t},$$

showing that (on average) amoebas multiply at an exponential rate (whereas a Poisson process grows linearly on average—remember (8)). The same type of argument may be used to calculate $E(M_t^2)$. This is more complicated and leads to an expression for the variance of M_t,

(37)
$$\text{var}(M_t) = Ie^{2\lambda t}(1 - e^{-\lambda t}).$$

An alternative method of calculating the mean and variance of M_t proceeds by way of the differential equation (33) for the probability generating function $G(s, t)$ of M_t. Remember that

$$G(1, t) = 1, \quad \left[\frac{\partial G}{\partial s}\right]_{s=1} = \mu(t).$$

We differentiate throughout (33) with respect to s and substitute $s = 1$ to obtain

$$\left[\frac{\partial^2 G}{\partial s\, \partial t}\right]_{s=1} = \left[\lambda \frac{\partial G}{\partial s}\right]_{s=1}.$$

Assuming that we may interchange the order of differentiation in the first term here, this equation becomes

$$\mu'(t) = \lambda\mu(t),$$

which is easily solved to yield $\mu(t) = Ie^{\lambda t}$. The variance may be found similarly, by differentiating twice.

Exercises 10. Show that, in the simple birth process above, the period of time during which there are exactly k ($\geq I$) individuals is a random variable having the exponential distribution with parameter λk.

11. Deduce from the result of Exercise 10 that the time T_{IJ} required by the birth process to grow from size I to size J ($> I$) has mean and variance given by

$$E(T_{IJ}) = \sum_{k=I}^{J-1} \frac{1}{\lambda k},$$

$$\text{var}(T_{IJ}) = \sum_{k=I}^{J-1} \frac{1}{(\lambda k)^2}.$$

12. Show that the variance of the simple birth process M_t is given by

$$\text{var}(M_t) = Ie^{2\lambda t}(1 - e^{-\lambda t}).$$

11.5   Birth and death processes

It is usually the case that in any system involving births, there are deaths also. Most telephone calls last for only a finite time, and most bacteria die out after their phases of self-reproduction. We introduce death into the simple birth process of the last section by replacing postulate (25), which concerned divisions, by two postulates concerning divisions and deaths respectively. We shall suppose that our hypothetical type of amoeba satisfies the following:

(38)   *division rate*: each amoeba present in the pond at time t has a chance of dividing in the short time-interval $(t, t + h]$; there exists a constant λ (>0), called the *birth rate*, such that the probability that any such amoeba
   (i) divides *once* during the time-interval $(t, t + h]$ equals $\lambda h + o(h)$,
   (ii) divides more than once during the time-interval $(t, t + h]$ equals $o(h)$,

(39)   *death rate*: each amoeba present at time t has a chance of dying, and hence being removed from the population, during the short time-interval $(t, t + h]$; there exists a constant μ (>0) called the *death rate*, such that the probability that any such amoeba dies during the time-interval $(t, t + h]$ equals $\mu h + o(h)$.

We assume also that deaths occur independently of other deaths and of all births. For the time-interval $(t, t + h]$ there are now several possibilities for each amoeba present at time t:
   (i) death, with probability $\mu h + o(h)$,
   (ii) a single division, with probability $\lambda h + o(h)$,
   (iii) no change of state, with probability $[1 - \lambda h + o(h)][1 - \mu h + o(h)] = 1 - (\lambda + \mu)h + o(h)$ (compare this with (25ii),
   (iv) some other combination of birth and death (such as division and death, or two divisions), with probability $o(h)$.
Only possibilities (i)–(iii) have probabilities which are large enough to be taken into account. Similarly, the probability of two or more amoebas changing their states (by division or death) during the time-interval $(t, t + h]$ equals $o(h)$. We write L_t for the number of amoebas which are alive at time t, and we find the distribution of L_t in the same way as before. The first step mimics (27)–(29), and similar calculations to those equations give that, for $k = 0, 1, 2, \ldots$,

(40) $$P(L_{t+h} = k \mid L_t = k) = 1 - (\lambda + \mu)kh + o(h),$$

(41) $$P(L_{t+h} = k + 1 \mid L_t = k) = \lambda k h + o(h),$$

(42) $$P(L_{t+h} = k - 1 \mid L_t = k) = \mu k h + o(h),$$

(43) $$P(L_{t+h} > k + 1 \text{ or } L_{t+h} < k - 1 \mid L_t = k) = o(h).$$

Note that if $L_t = k$ then the rate of birth in the population is λk and the rate of death is μk; this linearity in k arises because there are k independent possibilities for a birth or a death—remember (28).

This *birth–death process* differs from the simple birth process in a very important respect—it has an *absorbing state*, in the sense that if at some time there are no living cells, then there will never be any living cells subsequently.

Unlike the Poisson process and the simple birth process, it is not very easy to write down the mass function of L_t explicitly, since the corresponding differential–difference equations are not easily solved by recursion. The method of generating functions is still useful.

Theorem 11E

If $L_0 = I$ then L_t has probability generating function

(44)
$$E(s^{L_t}) = \begin{cases} \left(\dfrac{\lambda t(1-s)+s}{\lambda t(1-s)+1}\right)^I & \text{if } \mu = \lambda, \\[3mm] \left(\dfrac{\mu(1-s)-(\mu-\lambda s)e^{t(\mu-\lambda)}}{\lambda(1-s)-(\mu-\lambda s)e^{t(\mu-\lambda)}}\right)^I & \text{if } \mu \neq \lambda. \end{cases}$$

Proof

The differential–difference equations for $p_k(t) = P(L_t = k)$ are easily seen to be

(45)
$$p_k'(t) = \lambda(k-1)p_{k-1}(t) - (\lambda + \mu)kp_k(t) + \mu(k+1)p_{k+1}(t),$$

valid for $k = 0, 1, 2, \ldots$ subject to the convention that $p_{-1}(t) = 0$ for all t; the boundary condition is

(46)
$$p_k(0) = \begin{cases} 1 & \text{if } k = I, \\ 0 & \text{if } k \neq I. \end{cases}$$

Recursive solution of (45) fails since the equation in $p_0'(t)$ involves both $p_0(t)$ and $p_1(t)$ on the right-hand side. We solve these equations by the method of generating functions, first introducing the probability generating function

$$G(s, t) = E(s^{L_t}) = \sum_{k=0}^{\infty} p_k(t)s^k.$$

Multiply throughout (45) by s^k and sum over k to obtain the partial differential equation

$$\frac{\partial G}{\partial t} = (\lambda s - \mu)(s - 1)\frac{\partial G}{\partial s},$$

with boundary condition $G(s, 0) = s^I$. The diligent reader may check that the solution is given by (44). $\qquad\square$

It is possible that this birth–death process L will become extinct ultimately, in that $L_t = 0$ for some t. The probability that this happens is easily calculated from the result of Theorem 11E.

Theorem 11F

Let $L_0 = I$, and write $e(t)$ for the probability $P(L_t = 0)$ that the process is extinct by time t. Then, as $t \to \infty$,

(47)

$$e(t) \to \begin{cases} 1 & \text{if } \lambda \leq \mu, \\ (\mu/\lambda)^I & \text{if } \lambda > \mu. \end{cases}$$

Hence, extinction is certain if and only if the death rate is at least as big as the birth rate. This result is strikingly reminiscent of the Gambler's Ruin problem and Theorem 10E. Actually, (47) may be derived directly from the conclusion of Theorem 10E by studying what is called the 'embedded random walk' of the birth–death process (see Grimmett and Stirzaker (1982, p. 160) for the details).

Proof

Clearly

$$e(t) = P(L_t = 0) = G(0, t).$$

From (44),

$$G(0, t) = \begin{cases} \left(\dfrac{\lambda t}{\lambda t + 1}\right)^I & \text{if } \lambda = \mu, \\ \left(\dfrac{\mu - \mu e^{t(\mu - \lambda)}}{\lambda - \mu e^{t(\mu - \lambda)}}\right)^I & \text{if } \lambda \neq \mu, \end{cases}$$

and the result follows immediately. $\square$

Exercises

13. Let $m(t)$ be the expected size of the population in a birth–death process with birth rate λ and death rate μ. Use (45) to show that m satisfies the differential equation

$$m'(t) = (\lambda - \mu)m(t).$$

Hence find $m(t)$ in terms of the initial size of the population.

14. A birth–death process L has birth rate λ and death rate μ. If the population has size k at time t, show that the subsequent length of time which elapses before there is either a birth or a death is a random variable having the exponential distribution with parameter $(\lambda + \mu)k$.

15. Let L be a birth–death process with birth rate 1 and death rate 1. Suppose that L_0 is a random variable having the Poisson distribution with parameter α. Show that the probability that the process is extinct by time t is $\exp[-\alpha/(t + 1)]$.

11.6 A simple queueing model

We all know how it feels to be waiting in a queue, whether it be buying postage stamps at the post office at lunchtime, telephoning into a busy switchboard, waiting for a response from an on-line computing system, or waiting to be called in for a minor operation at the local hospital.

There are many different types of queue, and there are three principal ways in which we may categorize them, according to the ways in which

(i) people arrive in the system,

(ii) these people are stored in the system prior to their service,

(iii) these people are served.

In many queues, only the method (ii) of storage of waiting customers can be predicted with certainty—*first come, first served* is a common 'queue discipline' in shops, although there are many other possibilities. On the other hand, it is generally impossible to predict exactly when people will join the queue and how long they will require for service, and this is the reason why probability theory is important in describing queueing systems. The theory of queues is an old favourite amongst probabilists, being a rich source of interesting and diverse problems.

We shall consider a simple model of a queue; there are many others, most of which are more complicated than this one, and the reader may find amusement in devising some of these. Our example goes as follows. In the German delicatessen in the market, Eva is the only shop assistant. Customers arrive in the shop, wait for their turn to be served by Eva, and then leave after their service has been completed. There is randomness in the way in which they arrive and in the lengths of their service times (people rarely visit good delicatessens and buy only one type of cheese). We suppose that

(48)   *arrivals*: customers arrive in the manner of a Poisson process with rate λ (>0); that is to say, if N_t is the number who have arrived by time t, then $N = (N_t : t \geq 0)$ is a Poisson process with rate λ,

(49)   *service*: the service time of each customer is a random variable having the exponential distribution with parameter μ (>0), and the service times of different customers are independent random variables,

(50)   *independence*: service times are independent of arrival times, so that, for example, Eva works no faster when the shop is crowded than she does when it is nearly empty.

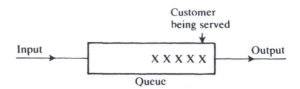

Fig. 11.2 A simple queue

It is not very important to us how a customer is stored between his arrival and his departure, but for the sake of definiteness we shall suppose that Eva tolerates no disorder, insisting that her customers form a single line and always serving the customer at the head of this line in the usual way; this queue discipline is called *first come, first served* or *first in, first out*. Thus, Eva's shop looks something like Fig. 11.2. Other delicatessens are less disciplined, with people milling around the shop and the shop assistant serving people chosen at random.

Assumption (48) is equivalent to demanding that the times between successive arrivals are independent exponentially distributed random variables with parameter λ. Our assumption that both inter-arrival times and service times have the exponential distribution is crucial for this example, since this is the unique distribution with the lack-of-memory property (Theorem 11C). This assumption has the following consequence. If we glance through the shop window at a certain time, seeing ten people within, say, then the times of the next arrival and the next departure do not depend on the times of the last arrival and the last departure. Thus, for example, for $h > 0$,

(51) the probability of a single arrival during the time-interval $(t, t+h]$ equals $\lambda h + o(h)$, and the probability of no arrival equals $1 - \lambda h + o(h)$.

Also, if Eva is serving someone at time t, then the probability that she is still serving this person at time $t+h$ equals $P(S > t + h - \tau \mid S > t - \tau)$, where S is the service time of the customer in question and τ is the time at which his service began; however,

$$P(S > t + h - \tau \mid S > t - \tau) = P(S > h)$$
$$= e^{-\mu h}$$
$$= 1 - \mu h + o(h),$$

by Theorem 11C and the lack-of-memory property (20), giving that, for $h > 0$,

(52)   if Eva is serving someone at time t, then the probability that this service is completed during the time-interval $(t, t+h]$ equals $\mu h + o(h)$.

Furthermore, if the shop is occupied at time t, then the events in (51) and (52) are independent.

Let Q_t be the number of people in the queue (including the person being served) at time t, and suppose that $Q_0 = 0$. The process $Q = (Q_t : t \geq 0)$ is a type of birth–death process since, if $Q_t = k$, say, then Q_{t+h} equals $k-1$, k, or $k+1$ with probability $1 - o(h)$; the only events which may happen with significant probability (that is, larger than $o(h)$) during the time-interval $(t, t+h]$ are a single departure, a single arrival, or no change of state. More precisely, if $k \geq 1$,

(53) $$P(Q_{t+h} = k \mid Q_t = k) = P(\text{no arrival, no departure}) + o(h)$$
$$= [1 - \lambda h + o(h)][1 - \mu h + o(h)] + o(h)$$
$$= 1 - (\lambda + \mu)h + o(h),$$

(54) $$P(Q_{t+h} = k-1 \mid Q_t = k) = P(\text{no arrival, one departure}) + o(h)$$
$$= [1 - \lambda h + o(h)][\mu h + o(h)] + o(h)$$
$$= \mu h + o(h),$$

and if $k \geq 0$,

(55) $$P(Q_{t+h} = k+1 \mid Q_t = k) = P(\text{one arrival, no departure}) + o(h)$$
$$= [\lambda h + o(h)][1 - \mu h + o(h)] + o(h)$$
$$= \lambda h + o(h).$$

Finally,

(56) $$P(Q_{t+h} = 0 \mid Q_t = 0) = P(\text{no arrival}) + o(h)$$
$$= 1 - \lambda h + o(h).$$

These equations (53)–(56) are very similar to the corresponding relations (40)–(42) for the simple birth–death process, the only significant difference being that arrivals and departures occur here at rates which do not depend upon the current queue size (unless it is empty, so that departures are impossible), whereas in the simple birth–death process the corresponding rates are linear functions of the current population size. It may seem at first sight that this queueing process is simpler than the birth–death process, but the truth turns out to be the opposite: there is no primitive way of calculating the mass function of $Q(t)$ if $t > 0$. The difficulty is as follows. As usual, we may use (53)–(56) to establish a system of differential–difference equations for the probabilities

$$p_k(t) = P(Q_t = k),$$

and these turn out to be

(57) $$p_k'(t) = \lambda p_{k-1}(t) - (\lambda + \mu)p_k(t) + \mu p_{k+1}(t) \qquad \text{for } k = 1, 2, \ldots ,$$

and

(58) $$p_0'(t) = -\lambda p_0(t) + \mu p_1(t),$$

subject to the boundary condition

$$p_k(0) = \begin{cases} 1 & \text{if } k = 0, \\ 0 & \text{otherwise.} \end{cases}$$

We cannot solve this system of equations recursively since the equation (58) for $p_0(t)$ involves $p_1(t)$ also. Furthermore the method of generating functions leads to the differential equation

$$\frac{\partial G}{\partial t} = \frac{s-1}{s}[\lambda s G - \mu G + \mu p_0(t)]$$

for $G(s, t) = E(s^{Q_t})$, and this equation involves the unknown function $p_0(t)$. Laplace transforms turn out to be the key to solving (57) and (58) and the answer is not particularly pretty:

(59) $$p_k(t) = J_k(t) - J_{k+1}(t) \qquad \text{for } k = 0, 1, 2, \ldots ,$$

where

$$J_k(t) = \int_0^t \left(\frac{\lambda}{\mu}\right)^{\frac{1}{2}k} \frac{k}{s} e^{-s(\lambda+\mu)} I_k(2s\sqrt(\lambda\mu)) \, ds$$

and $I_k(t)$ is a modified Bessel function. We shall not prove this here, of course, but refer those interested to Feller (1971, p. 482).

The long-term behaviour of the queue is of major interest. If service times are long in comparison with inter-arrival times then the queue will tend to grow, so that after a long period of time it will be very large; on the other hand, if service times are relatively short then it is reasonable to expect that the queue length will settle down into some 'steady state'. The asymptotic behaviour of the queue length Q_t as $t \to \infty$ is described by the sequence

(60) $$\pi_k = \lim_{t \to \infty} p_k(t) \qquad \text{for } k = 0, 1, 2, \ldots ,$$

of limiting probabilities, and it is in this sequence $\pi_0, \pi_1, \ldots$ that we are interested. It is in fact the case that the limits exist in (60), but we shall not prove this. Neither do we prove that

(61) $$0 = \lim_{t \to \infty} p_k'(t) \qquad \text{for } k = 0, 1, 2, \ldots ,$$

a fact which follows intuitively from (60), by differentiating both sides of (60) and interchanging the limit and the differential operator. The values of the π's are found by letting $t \to \infty$ in (57) and (58) and using (60) and (61) to obtain the following difference equations for the sequence $\pi_0, \pi_1, \ldots$:

(62) $$0 = \lambda \pi_{k-1} - (\lambda + \mu)\pi_k + \mu \pi_{k+1} \qquad \text{for } k = 1, 2, \ldots ,$$

(63) $$0 = -\lambda \pi_0 + \mu \pi_1.$$

We call a non-negative sequence $\pi_0, \pi_1, \ldots$ a *steady-state distribution* of the queue if it satisfies (62) and (63) together with

(64) $$\sum_{k=0}^{\infty} \pi_k = 1.$$

Theorem 11G   *If $\lambda < \mu$ then the queue has a steady-state distribution given by*

(65) $$\pi_k = \left(1 - \frac{\lambda}{\mu}\right)\left(\frac{\lambda}{\mu}\right)^k \qquad \text{for } k = 0, 1, 2, \ldots .$$

If $\lambda \geq \mu$, there is no steady-state distribution.

We call the ratio $\rho = \lambda/\mu$ the *traffic intensity* of the queue; ρ is the ratio of the arrival rate to the service rate. We may interpret Theorem 11G as follows:

(a) if $\rho < 1$, the queue length Q_t settles down as $t \to \infty$ into a steady-state or 'equilibrium' distribution, for which the probability that k customers are present equals $(1 - \rho)\rho^k$;

(b) if $\rho \geq 1$, there is no steady-state distribution, indicating that the rate of arrival of new customers is too large for the single server to cope, and the queue length either grows beyond all bounds or it has fluctuations of a very large order.

Theorem 11G may remind the reader of the final theorem (Theorem 10E) about a random walk with an absorbing barrier. Just as in the case of the simple birth–death process, there is a random walk embedded in this queueing process, and this random walk has a 'retaining' barrier at 0 which prevents the walk from visiting the negative integers but allows the walk to re-visit the positive integers.

Proof of Theorem 11G   We wish to solve the difference equations (62) and (63), and we do this recursively rather than using the general method of the appendix. We find π_1 in terms of π_0 from (63):

$$\pi_1 = \rho \pi_0.$$

We substitute this into (62) with $k = 1$ to find that

$$\pi_2 = \rho^2 \pi_0,$$

and we deduce the general solution

(66) $\pi_k = \rho^k \pi_0$ for $k = 0, 1, 2, \ldots$

by induction on k. Now, $\pi_0, \pi_1, \ldots$ is a steady-state solution if and only if (64) holds. From (66),

$$\sum_{k=0}^{\infty} \pi_k = \pi_0 \sum_{k=0}^{\infty} \rho^k.$$

If $\rho < 1$ then (64) holds if and only if $\pi_0 = 1 - \rho$. On the other hand if $\rho \geq 1$ then (64) holds for no value of π_0, and the proof is complete.

□

Exercises 16. If Q is the above queueing process with arrival rate λ and service rate μ, and a customer arrives to find exactly k customers waiting ahead of him (including the person being served), show that this customer leaves the queueing system after a length of time which has the gamma distribution with parameters $k + 1$ and μ.
17. Show that $p_k(t) = (1 - \rho)\rho^k$ is a solution to equations (57) and (58) so long as $\rho = \lambda/\mu < 1$. This proves that if the process begins in its steady-state distribution then it has this distribution for all time.
18. A queue has three servers A_1, A_2, A_3, and their service times are independent random variables, A_i's service times having the exponential distribution with parameter μ_i. A customer arrives and finds all three servers unoccupied; he chooses one at random, each being equally likely. If he is still being served at time t, what is the probability that he chose A_1?

11.7 Problems

1. If N is a Poisson process with rate λ, what is the distribution of $N_t - N_s$ for $0 \leq s \leq t$?
2. If N is a Poisson process with rate λ, show that $\text{cov}(N_s, N_t) = \lambda s$ if $0 \leq s < t$.
3. Three apparently identical robots, called James, Simon, and John, are set to work at time $t = 0$. Subsequently each stops working after a random length of time, independently of the other two, and the probability that any given robot stops in the short time-interval $(t, t + h)$ is $\lambda h + o(h)$. Show that each robot works for a period of time with the exponential distribution, parameter λ, and that the probability that at least one of the three has stopped by time t is $1 - e^{-3\lambda t}$.
What is the probability that they stop work in the order James, Simon, John?

*4. Let $X_1, X_2, X_3, \ldots$ be a sequence of independent identically distributed random variables having the distribution function

$$F(x) = \begin{cases} 1 - e^{-\lambda x} & (x \geq 0), \\ 0 & (x < 0), \end{cases}$$

where λ is a positive constant. If $S_n = \sum_{i=1}^n X_i$ then prove that S_n has density function

$$f_n(x) = \lambda^n x^{n-1} e^{-\lambda x}/(n-1)! \qquad (x \geq 0).$$

Deduce that $N_t = \max\{n : S_n \leq t\}$ has a Poisson distribution. The *excess life e_t* is defined by

$$e_t = S_{N_t+1} - t.$$

If $g(t, x) = P(e_t > x)$ then by considering the distribution of e_t conditional on the value of X_1 show that

$$g(t, x) = e^{-\lambda(t+x)} + \int_0^t g(t - u, x)\lambda e^{-\lambda u}\, du.$$

Find a solution of this equation. (Oxford 1976F)

5. Tourist coaches arrive at Buckingham Palace in the manner of a Poisson process with rate λ, and the numbers of tourists in the coaches are independent random variables each having probability generating function $G(s)$. Show that the total number of tourists who have arrived at the palace by time t has probability generating function

$$\exp\{\lambda t[G(s) - 1]\}.$$

This is an example of a so-called 'compound' Poisson process.

6. The probability of one failure in a system occurring in the time interval $(t, t + \tau)$ is $\lambda(t)\tau + o(\tau)$, independently of previous failures, and the probability of more than one failure in this interval is $o(\tau)$, where λ is a positive integrable function.
Prove that the number of failures in the interval $(0, t)$ has a Poisson distribution with mean $\int_0^t \lambda(x)\, dx$.
Let T be the time of occurrence of the first failure. Find the probability density function of T and show that, if $\lambda(t) = c/(1 + t)$ (where $c > 0$), the expected value of T is finite if and only if $c > 1$. (Oxford 1981F)
This is an example of a so-called 'inhomogeneous' Poisson process.

7. A 'doubly stochastic' Poisson process is an inhomogeneous Poisson process in which the rate function $\lambda(t)$ is itself a random process. Show that the simple birth process with birth rate λ is a doubly stochastic Poisson process N in which $\lambda(t) = \lambda N_t$.

8. In a simple birth process with birth rate λ, find the moment generating function of the time required by the process to grow from size I to size J ($> I$).

9. Show that the moment generating function of the so-called 'extreme value' distribution with density function

$$f(x) = \exp(-x - e^{-x}) \qquad \text{for } x \in \mathbb{R},$$

is

$$M(t) = \Gamma(1 - t) \qquad \text{if } t < 1.$$

Let T_J be the time required by a simple birth process with birth rate λ to grow from size 1 to size J, and let

$$U_J = \lambda T_J - \log J.$$

Show that U_J has moment generating function

$$M_J(t) = \frac{1}{J^t} \prod_{i=1}^{J-1} \left(\frac{i}{i-t}\right) \qquad \text{if } t < 1,$$

and deduce that, as $J \to \infty$,

$$M_J(t) \to M(t) \qquad \text{if } t < 1.$$

(You may use the fact that $J^t \Gamma(J-t)/\Gamma(J) \to 1$ as $J \to \infty$.) It follows that the distribution of U_J approaches the extreme-value distribution as $J \to \infty$.

10. Consider a birth–death process whose birth and death rates satisfy $\lambda = \mu$. If the initial population size is I, show that the time T until the extinction of the process has distribution function

$$P(T \le t) = \left(\frac{\lambda t}{\lambda t + 1}\right)^I \qquad \text{for } t > 0,$$

and deduce that, as $I \to \infty$, the random variable $U_I = \lambda T / I$ has limiting distribution function given by

$$P(U_I \le t) \to e^{-1/t} \qquad \text{for } t \ge 0.$$

11. A population develops according to the following rules: (i) during the interval $(t, t + dt)$ an individual existing at time t has (independently of its previous history) probability $\lambda \, dt + o(dt)$ of having a single offspring (twins, triplets, etc. being impossible) and a probability $\mu \, dt + o(dt)$ of dying, where λ and μ are absolute constants; (ii) in the interval $(t, t + dt)$ there is a probability $\theta \, dt + o(dt)$ that a single immigrant will join the population; and (iii) the subpopulations descending from distinct individuals develop independently. If $p_n(t)$ denotes the probability that the population consists of n individuals at time t, show that

$$\phi(z, t) \equiv \sum_{n=0}^{\infty} z^n p_n(t)$$

satisfies the partial differential equation

$$\frac{\partial \phi}{\partial t} = (\lambda z - \mu)(z - 1)\frac{\partial \phi}{\partial z} + \theta(z - 1)\phi.$$

In the particular case when $\lambda = \mu = \theta = 1$ and the population is empty at time $t = 0$, show that the size of the population at time t has mean t and calculate its variance. (Oxford 1964F)

12. Consider the 'birth–death–immigration' process of Problem 11 and suppose that λ, μ, $\theta > 0$. Use the ideas and methods of Section 11.6 to show that this process has a steady-state distribution if and only if $\lambda < \mu$, and in this case the steady-state distribution is given by

$$\pi_n = \pi_0 \frac{1}{n!} \left(\frac{\lambda}{\mu}\right)^n \frac{\Gamma(n + (\theta/\lambda))}{\Gamma(\theta/\lambda)} \qquad \text{for } n = 0, 1, 2, \ldots,$$

where π_0 is chosen so that $\sum_n \pi_n = 1$.

13. The 'immigration–death' process is obtained from the birth–death–immigration process of Problem 11 by setting the birth rate λ equal to 0. Let $D = (D_t : t \geq 0)$ be an immigration–death process with positive immigration rate θ and death rate μ. Suppose that $D_0 = I$, and set up the system of differential equations which are satisfied by the probability functions

$$p_k(t) = P(D_t = k).$$

Deduce that the probability generating function

$$G(s, t) = E(s^{D_t})$$

of D_t satisfies the partial differential equation

$$\frac{\partial G}{\partial t} = (s - 1)\left(\theta G - \mu \frac{\partial G}{\partial s}\right)$$

subject to the boundary condition $G(s, 0) = s^I$. Solve this equation to find that

$$G(s, t) = [1 + (s - 1)e^{-\mu t}]^I \exp[\theta(s - 1)(1 - e^{-\mu t})/\mu].$$

14. In the immigration–death process of Problem 13, show that there is a steady-state distribution (in the jargon of Section 11.6) for all positive values of θ and μ. Show further that this distribution is the Poisson distribution with parameter θ/μ.

15. A robot can be in either of two states: state A (idle) and state B (working). In any short time-interval $(t, t + h)$, the probability that it changes its state is $\lambda h + o(h)$ where $\lambda > 0$. If $p(t)$ is the probability that it is idle at time t given that it was idle at time 0, show that

$$p'(t) = -2\lambda p(t) + \lambda.$$

Hence find $p(t)$.

Let $q(t)$ be the probability that the robot is working at time t given that it was working at time 0. Find $q(t)$, and find the distribution of the earliest time T at which there is a change of state.

If there are N robots operating independently according to the above laws and $p_k(t)$ is the probability that exactly k are idle at time t, show that

$$p_k'(t) = \lambda(N - k + 1)p_{k-1}(t) - \lambda N p_k(t) + \lambda(k + 1)p_{k+1}(t),$$

for $k = 0, 1, \ldots, N$, subject to the rule that $p_{-1}(t) = p_{N+1}(t) = 0$.
If all the robots are idle at time 0, show that the number of idle robots at time t has the binomial distribution with parameters N and $e^{-\lambda t} \cosh(\lambda t)$.

16. Prove that, in a queue whose input is a Poisson process and whose service times have the exponential distribution, the number of new arrivals during any given service time is a random variable with the geometric distribution.

17. Customers arrive in a queue according to a Poisson process with rate λ and their service times have the exponential distribution with parameter μ. Show that if there is only one customer in the queue then the probability that the next customer arrives within time t and has to wait for service is

$$\frac{\lambda}{\lambda + \mu}(1 - e^{-(\lambda + \mu)t}).$$

18. Customers arrive in a queue according to a Poisson process with rate λ and their service times have the exponential distribution with parameter μ, where $\lambda < \mu$. We suppose that the number Q_0 of customers in the system at time 0 has distribution

$$P(Q_0 = k) = (1 - \rho)\rho^k \qquad \text{for } k = 0, 1, 2, \ldots,$$

where $\rho = \lambda/\mu$, so that the queue is 'in equilibrium' by the conclusion of Exercise 17 in Section 11.6. If a customer arrives in the queue at time t, find the moment generating function of the total time which he spends in the system, including his service time. Deduce that this time has the exponential distribution with parameter $\mu(1 - \rho)$.

19. Customers arrive at the door of a shop according to a Poisson process with rate λ, but they are unwilling to enter a crowded shop. If a prospective customer sees k people inside the shop then he enters the shop with probability $(\frac{1}{2})^k$ and otherwise leaves, never to return. The service times of customers who enter the shop are random variables with the exponential distribution, parameter μ. If Q_t is the number of people within the shop (excluding the single server) at time t, show that the probability functions

$$p_k(t) = P(Q_t = k)$$

satisfy

$$p_k'(t) = \mu p_{k+1}(t) - \left(\frac{\lambda}{2^k} + \mu\right)p_k(t) + \frac{\lambda}{2^{k-1}}p_{k-1}(t)$$

for $k = 1, 2, \ldots$, and

$$p_0'(t) = \mu p_1(t) - \lambda p_0(t).$$

Deduce that there is a steady-state distribution for all positive values of λ and μ, and that this distribution is given by

$$\pi_k = \pi_0 2^{-\frac{1}{2}k(k-1)}\rho^k \qquad \text{for } k = 0, 1, 2, \ldots,$$

where $\rho = \lambda/\mu$ and π_0 is chosen appropriately.

Appendix
Difference equations

We say that the sequence $x_0, x_1, \ldots$ satisfies a *difference equation* if

(1)
$$a_0 x_{n+k} + a_1 x_{n+k-1} + \cdots + a_k x_n = 0 \qquad \text{for } n = 0, 1, 2, \ldots,$$

where $a_0, a_1, \ldots, a_k$ is a given sequence of real numbers and $a_0 \neq 0$. We generally suppose that $a_k \neq 0$, and in this case we call (1) a difference equation of *order k*. Difference equations occur quite often in studying random processes, particularly random walks, and it is useful to be able to solve them. We describe here how to do this.

Just as in solving differential equations, we require boundary conditions in order to solve difference equations. To see this, note that (1) may be rewritten as

(2)
$$x_{n+k} = -\frac{1}{a_0}(a_1 x_{n+k-1} + a_2 x_{n+k-2} + \cdots + a_k x_n) \qquad \text{for } n = 0, 1, 2, \ldots$$

since $a_0 \neq 0$. Thus, if we know the values of $x_0, x_1, \ldots, x_{k-1}$, equation (2) with $n = 0$ provides the value of x_k. Next, equation (2) with $n = 1$ tells us the value of x_{k+1}, and so on. That is to say, there is a unique solution of (1) with specified values for $x_0, x_1, \ldots, x_{k-1}$. It follows that, if $a_k \neq 0$, the general solution of (1) contains exactly k independent arbitrary constants, and so exactly k independent boundary conditions are required in order to solve (1) explicitly.

The principal step involved in solving (1) is to find the roots of the *auxiliary equation*

(3)
$$a_0 \theta^k + a_1 \theta^{k-1} + \cdots + a_{k-1}\theta + a_k = 0,$$

a polynomial in θ of degree k. We denote the (possibly complex) roots of this polynomial by $\theta_1, \theta_2, \ldots, \theta_k$. The general solution of (1) is given in the next theorem.

Theorem A *Suppose that $a_0, a_1, \ldots, a_k$ is a given sequence of real numbers and $a_0 \neq 0$.*

(i) If the roots $\theta_1, \theta_2, \ldots, \theta_k$ of the auxiliary equation are distinct, the general solution of (1) is

(4)
$$x_n = c_1 \theta_1^n + c_2 \theta_2^n + \cdots + c_k \theta_k^n \qquad \text{for } n = 0, 1, 2, \ldots,$$

where $c_1, c_2, \ldots, c_k$ are arbitrary constants.

(ii) *More generally, if $\theta_1, \theta_2, \ldots, \theta_r$ are the distinct roots of the auxiliary equation and m_i is the multiplicity of θ_i for $i = 1, 2, \ldots, r$, the general solution of (1) is*

(5)
$$x_n = (a_1 + a_2 n + \cdots + a_{m_1} n^{m_1-1})\theta_1^n$$
$$+ (b_1 + b_2 n + \cdots + b_{m_2} n^{m_2-1})\theta_2^n + \cdots$$
$$+ (c_1 + c_2 n + \cdots + c_{m_r} n^{m_r-1})\theta_r^n \quad \text{for } n = 0, 1, 2, \ldots,$$

where the k numbers $a_1, \ldots, a_{m_1}, b_1, \ldots, b_{m_2}, \ldots, c_1, \ldots, c_{m_r}$ are arbitrary constants.

The auxiliary equation may not possess k *real* roots, and thus some or all of $\theta_1, \theta_2, \ldots, \theta_k$ may have non-zero imaginary parts. Similarly the arbitrary constants in Theorem A need not necessarily be real, and the general solution (5) is actually the general solution for *complex* solutions of the difference equation (1). If we seek *real* solutions only of (1), then this fact should be taken into account when finding the values of the constants.

We do not prove this theorem, but here are two ways of going about proving it, should one wish to do so. The first way is constructive, and uses the generating functions of the sequences of a's and x's (see Hall (1967, p. 20)). The second way is to check that (4) and (5) are indeed solutions of (1) and then to note that they contain the correct number of arbitrary constants.

Here is an example of the theorem in action.

Example 6 Find the solution of the difference equation

$$x_{n+3} - 5x_{n+2} + 8x_{n+1} - 4x_n = 0$$

subject to the boundary conditions $x_0 = 0$, $x_1 = 3$, $x_3 = 41$.

Solution The auxiliary equation is

$$\theta^3 - 5\theta^2 + 8\theta - 4 = 0$$

with roots $\theta = 1, 2, 2$. The general solution is therefore

$$x_n = a 1^n + (b + cn)2^n$$

where the constants a, b, c are found from the boundary conditions to be given by $a = 1$, $b = -1$, $c = 2$. $\square$

An important generalization of Theorem A deals with difference equations of the form

(7)
$$a_0 x_{n+k} + a_1 x_{n+k-1} + \cdots + a_k x_n = g(n) \quad \text{for } n = 0, 1, 2, \ldots,$$

where g is a given function of n, not always equal to 0. There are two principal steps in solving (7). First we find a solution of (7) by any means available, and we call this a *particular solution*. Secondly, we find the general solution to the difference equation obtained by setting $g(n) = 0$ for all n:

$$a_0 x_{n+k} + a_1 x_{n+k-1} + \cdots + a_k x_n = 0 \qquad \text{for } n = 0, 1, 2, \ldots,$$

this solution is called the *complementary solution*.

Theorem B *Suppose that $a_0, a_1, \ldots, a_k$ is a given sequence of real numbers and $a_0 \neq 0$. The general solution of (7) is*

(8)
$$x_n = \kappa_n + \pi_n \qquad \text{for } n = 0, 1, 2, \ldots,$$

where $\kappa_0, \kappa_1, \ldots$ is the complementary solution and $\pi_0, \pi_1, \ldots$ is a particular solution.

This may be proved in the same general way as Theorem A. We finish with an example.

Example 9 Find the solution of the difference equation

(10)
$$x_{n+2} - 5x_{n+1} + 6x_n = 4n + 2$$

subject to the boundary conditions $x_0 = 5$, $x_4 = -37$.

Solution The right-hand side of (10) is a polynomial function of n, and this suggests that there may be a particular solution which is a polynomial. Trial and error shows that

$$x_n = 2n + 4 \qquad \text{for } n = 0, 1, 2, \ldots$$

is a *particular solution*. The general solution of

$$x_{n+2} - 5x_{n+1} + 6x_n = 0$$

is

$$x_n = a2^n + b3^n \qquad \text{for } n = 0, 1, 2, \ldots,$$

where a and b are arbitrary constants. It follows that the general solution of (10) is

$$x_n = a2^n + b3^n + 2n + 4 \qquad \text{for } n = 0, 1, 2, \ldots.$$

The constants a and b are found from the boundary conditions to be given by $a = 2$, $b = -1$. □

Answers to exercises

Chapter 1

6. Yes.
9. $\frac{6}{10}$.
16. Compare $1 - (\frac{5}{6})^4$ with $1 - (\frac{35}{36})^{24}$.
19. $\frac{1}{8}$.
21. Either A or B must have zero probability.
24. (i) $(1-p)^m$ (ii) $\frac{1}{2}(1 + (q-p)^n)$, where $p + q = 1$.
26. $\frac{46}{63}$.
27. $\frac{4}{3}(\frac{2}{3})^n - \frac{1}{3}(-\frac{1}{3})^n$.

Chapter 2

4. Only V is a random variable.
5. $c = 1$.
8. $P(Y = 0) = e^{-\lambda} \cosh \lambda$, $P(Y = 1) = e^{-\lambda} \sinh \lambda$.
9. npq.

Chapter 3

1.

X \ Y	0	1	2
0	$\dfrac{11 \cdot 43}{13 \cdot 51}$	$\dfrac{88}{13 \cdot 51}$	$\dfrac{1}{13 \cdot 17}$
1	$\dfrac{88}{13 \cdot 51}$	$\dfrac{8}{13 \cdot 51}$	0
2	$\dfrac{1}{13 \cdot 17}$	0	0

2. $p_X(i) = \theta^i(\theta + \theta^2 + \theta^3)$ for $i = 0, 1, 2$.
7. Take $Y = X$, so that $X + Y$ takes even values only—it cannot then have the Poisson distribution.

Chapter 4

1. (i) $V(s) = 2U(s)$ (ii) $V(s) = U(s) + (1-s)^{-1}$ (iii) $V(s) = sU'(s)$.
2. $u_{2n} = \binom{2n}{n}p^n q^n$, $u_{2n+1} = 0$.

Chapter 5

2. Yes.

3. $F_Y(y) = \begin{cases} F_X(y) & \text{if } y \geq 0, \\ 0 & \text{if } y < 0. \end{cases}$

6. $c = \frac{1}{2}$.

7. $F(x) = \begin{cases} 0 & \text{if } x < 0, \\ x^2 & \text{if } 0 \leq x < 1, \\ 1 & \text{if } x \geq 1. \end{cases}$

8. $F(x) = \begin{cases} \frac{1}{2}e^x & \text{if } x \leq 0, \\ 1 - \frac{1}{2}e^{-x} & \text{if } x > 0. \end{cases}$

9. $f(x) = F'(x)$ if $x \neq 0$.

10. $F(x) = e^{-e^{-x}}$.

11. $w = 1$, arbitrary positive λ.

13. $2/\pi$.

14. (i) $f_A(x) = \frac{1}{2}\lambda \exp[-\frac{1}{2}\lambda(x - 5)]$ if $x > 5$.
    (ii) $f_B(x) = \lambda x^{-\lambda - 1}$ if $x > 1$.
    (iii) $f_C(x) = \lambda x^{-2} \exp[-\lambda(x^{-1} - 1)]$ if $x < 1$.
    (iv) $f_D(x) = \frac{1}{2}\lambda x^{-\frac{3}{2}} \exp[-\lambda(x^{-\frac{1}{2}} - 1)]$ if $x < 1$.

17. e^2.

Chapter 6

4. $c = \frac{6}{7}$ and $F(x, y) = \frac{6}{7}(\frac{1}{3}x^3 y + \frac{1}{8}x^2 y^2)$ if $0 \leq x \leq 1$, $0 \leq y \leq 2$.

5. $P(X + Y \leq 1) = 1 - 2e^{-1}$, $P(X > Y) = \frac{1}{2}$.

6. $c = 3$, $f_X(x) = 3x^2$ if $0 < x < 1$, and $f_Y(y) = \frac{3}{2}(1 - y^2)$ if $0 < y < 1$. X and Y are dependent.

7. X, Y, and Z are independent, and $P(X > Y) = P(Y > Z) = \frac{1}{2}$.

8. $f_{X+Y}(u) = \frac{1}{2}u^2 e^{-u}$ if $u > 0$.

10. $X + Y$ has the normal distribution with mean 0 and variance 2.

11. $f_{U,V}(u, v) = \dfrac{1}{4\pi\sigma^2} \exp\left(-\dfrac{1}{4\sigma^2}[u^2 + (v - 2\mu)^2]\right)$,

   which factorizes, so that U and V are independent.

13. $f_{X|Y}(x \mid y) = y^{-1}$ and $f_{Y|X}(y \mid x) = e^{x-y}$, if $0 < x < y < \infty$.

15. $E(\sqrt{(X^2 + Y^2)}) = \frac{2}{3}$ and $E(X^2 + Y^2) = \frac{1}{2}$.

16. Let X and Z be independent, X having the normal distribution with mean 0 and variance 1, and Z taking the values ± 1 each with probability $\frac{1}{2}$. Define $Y = XZ$.

17. $E(X \mid Y = y) = \frac{1}{2}y$ and $E(Y \mid X = x) = x + 1$.

Chapter 7

7. (i) $(\lambda/(\lambda - t))^w$ if $t < \lambda$ (ii) $\exp(-\lambda + \lambda e^t)$.

8. $\mu^3 + 3\mu\sigma^2$.

12. (i) $[\lambda/(\lambda - it)]^w$ (ii) $\exp(-\lambda + \lambda e^{it})$.

Chapter 8

8. $a = -\sqrt{6}$, $b = 2\sqrt{6}$.

Chapter 9

    2. $1 - p^n$.

Chapter 10

    1. $E(S_n) = n(p - q), \quad \text{var}(S_n) = 4pqn$.
    2. p^n.
    3. $\binom{2n + 1}{n} p^{n+1} q^n$.

Chapter 11

    1. $p(t) = e^{-\lambda t}$ if $t \geq 0$.
    9. The gamma distribution with parameters 2 and λ.
    13. $m(t) = m(0)e^{(\lambda - \mu)t}$.
    18. $e^{-\mu_1 t}(e^{-\mu_1 t} + e^{-\mu_2 t} + e^{-\mu_3 t})^{-1}$.

Remarks on the problems

Chapter 1

1. Expand $(1+x)^n + (1-x)^n$.
2. No.
5. $\frac{79}{140}$ and $\frac{40}{61}$.
6. $\frac{11}{50}$.
7. If X and Y are the numbers of heads obtained,

$$P(X = Y) = \sum_k P(X = k)P(Y = k) = \sum_k P(X = k)P(Y = n - k)$$

$$= P(X + Y = n).$$

8. $1 - (1-p)(1-p^2)^2$ and $1 - (1-p)(1-p^2)^2 - p + p[1 - (1-p)^2]^2$.
10. To do this rigorously is quite hard. You need only show that the proportion $\frac{1}{10}$ is correct for any single one of the numbers $0, 1, 2, \ldots, 9$.
11. Use the partition theorem 1B to obtain the difference equations. Either iterate these directly to solve them, or set up a matrix recurrence relation, and iterate this.
12. (a) Induction. (b) Let A_i be the event that the ith key is hung on its own hook. We see no advantage in using (a) for the first part.
13. Use the result of Problem 12(a).
14. Conditional probabilities again. The answer is $\frac{1}{4}(2e^{-1} + e^{-2} + e^{-4})$.
17. $\bigcup_{i=1}^n A_i \rightarrow \bigcup_{i=1}^\infty A_i$ as $n \rightarrow \infty$.
18. $\theta_n - (1 - \alpha - \beta)\theta_{n-1} = \alpha$. Such difference equations may be solved by the method described in the appendix: $\theta_n = (\theta_0 - \alpha(\alpha + \beta)^{-1})(1 - \beta - \alpha)^n + \alpha(\alpha + \beta)^{-1}$.

Chapter 2

2. Use Theorem 2C with X and B_i chosen appropriately. The answer is $m(r) = rp^{-1}$.
3. $E(X^2) = \sum x^2 P(X = x)$, the sum of non-negative terms.
4. $\alpha < -1$ and $c = \zeta(-\alpha)^{-1}$, where $\zeta(p) = \sum k^{-p}$ is the Riemann zeta function.
5. For the last part, show that $G(n) = P(X > n)$ satisfies $G(m + n) = G(m)G(n)$, and solve this relation.
6. The summation here is $\sum_{k=0}^\infty \sum_{i=k+1}^\infty P(X = i)$. Change the order of summation. For the second part, use the result of Exercise 8 of Section 1.4.
7. This generalizes the result of Problem 6.
8. This is sometimes called Banach's Matchbox Problem. First condition on which pocket is first emptied. Remember the negative binomial distribution. You may find the hint more comprehensible if you note that

$(n - h)p_h = \frac{1}{2}(2n + 1)p_{h+1} - \frac{1}{2}(h + 1)p_{h+1}$. The mean equals $(2n + 1)p_0 - 1$.

9. $(1 - p^n)[p^n(1 - p)]^{-1}$.

Chapter 3

1. Use the result of Exercise 18 of Section 1.6, with Theorem 3E.
2. $a = b = \frac{1}{2}$. No.
4. $P(U_n = k) = P(U_n \geq k) - P(U_n \geq k + 1)$, and $P(U_n \geq k) = \left(1 - \dfrac{k - 1}{N}\right)^n$.
5. $P(U > k) = P(X > k)P(Y > k)$.
6. Use Theorem 2C, with $B_i = \{N = i - 1\}$.
7. (a) $\frac{1}{2}$ (b) $\frac{1}{6}(3\sqrt{5} - 1)$ (c) $\frac{5}{6}$.
8. Use indicator functions as follows. Let Z_i be a random variable which equals 1 if the ith box is empty and 0 otherwise. The total number of empty boxes is $S = Z_1 + Z_2 + \cdots + Z_M$. Also $E(Z_i) = P(Z_i = 1) = (M - 1)^N/M^N$ and $E(S) = E(Z_1) + \cdots + E(Z_M)$.
9. Adapt the hint for Problem 8.
10. In calculating the mean, remember that the expectation operator E is linear. The answer here is $c\left(1 + \dfrac{1}{2} + \dfrac{1}{3} + \cdots + \dfrac{1}{c}\right)$, a much more elegant solution than that of Problem 7 in Section 2.6.
11. $c[1 - (1 - c^{-1})^n]$. This is a simple application of the method of indicator functions described above.
12. Condition on the value of N. X has the Poisson distribution with parameter λp.
13. $\text{var}(U_n) = (n - 1)q(1 - q) + (n - 2)(n - 3)q^2(1 - q)^2 - (n - 1)^2q^2(1 - q)^2$.

Chapter 4

1. Note that $P(X = k) = u_{k-1} - u_k$.
2. $6^{-7}\left(\dfrac{13!}{6! \, 7!} - 49\right)$.
3. $\frac{9}{19}, \frac{6}{19}, \frac{4}{19}$. Mean number of throws is 3.
4. $[q/(1 - ps)]^N$. The variance is $Np(1 - p)^{-2}$.
5. The first part of this problem may be done by way of Theorem 4D, with $N + 1$ having a geometric distribution and the X's having the Bernoulli distribution. Alternatively, use the methods of Chapter 3. The answer to the first part is $2(1 - p)p^r(2 - p)^{-r-1}$, and to the second part $\dbinom{n}{r}(\frac{1}{2})^{n+1}p^{n-r}(2 - p)^{r+1}$.
6. For the third part, find the real part of $G_X(\theta)$, where θ is a primitive complex cube root of unity.
8. $G_X(s) = G_N(\frac{1}{2} + \frac{1}{2}s)$, giving by independence that $G = G_N$ satisfies $G(s) = G(\frac{1}{2} + \frac{1}{2}s)^2$. Iterate this to obtain $G(s) = G(1 + (s - 1)/m)^m$ where $m = 2^n$, use Taylor's Theorem and take the limit as $n \to \infty$.
9. This is an alternative derivation of the result of Problem 10 in Section 3.5.

10. $P(A \text{ wins}) = a(a + b - ab)^{-1}$. The mean number of shots is $(2 - a)(a + b - ab)^{-1}$.

Chapter 5

3. $n \geq 4$.
4. $f_Y(y) = \sqrt{(2/\pi)}\exp(-\tfrac{1}{2}y^2)$ for $y > 0$. $\sqrt{(2/\pi)}$ and $1 - (2/\pi)$.
5. Let $F^{-1}(y) = \sup\{x : F(x) = y\}$. Find $P(F(X) \leq y)$.
6. Note that $x \leq F(y)$ if and only if $F^{-1}(x) \leq y$, for values of y satisfying $0 < F(y) < 1$.
7. Integrate by parts. You are proving that $E(X) = \int P(X > x)\, dx$, the continuous version of Problem 6 in Section 2.6.
8. Apply the conclusion of Problem 7 to $Y = g(X)$, express the result as a double integral and change the order of integration.
10. $f_Y(y) = \tfrac{5}{4}(1 + y)^{-2}$ for $\tfrac{1}{4} < y < \infty$.
11. This distance has distribution function $(2/\pi)\arctan x$ for $0 \leq x < \infty$.

Chapter 6

1. Find the joint density function of X/Y and XY by the method of change of variables, and then find the marginal density function of X.
2. No.
3. The region $\{(x, y, z) : \sqrt{(4xz)} < y \leq 1,\ 0 \leq x, z \leq 1\}$ has volume $\tfrac{5}{36} + \tfrac{1}{6}\log 2$.
4. $\text{Min}\{X, Y\} > u$ if and only if $X > u$ and $Y > u$.
5. Show that $G(y) = P(Y > y)$ satisfies $G(x + y) = G(x)G(y)$, and solve this equation.
6. If you can do Problem 4 then you should be able to do this one. $P(U \leq x, V \leq y) = F(y)^n - [F(y) - F(x)]^n$ for $x < y$.
9. $f_Y(y) = \tfrac{1}{3}(2y + 1)e^{-y}$ for $0 < y < \infty$. $F_{W,Z}(w, z) = F_{X,Y}(w, z) + F_{X,Y}(z, w) - F_{X,Y}(w, w)$.
10. S and $T - S$ each have the exponential distribution with parameter λ, and Z has density function $2\lambda e^{-\lambda u}(1 - e^{-\lambda u})^2$ for $u > 0$.
12. You need $\varepsilon > 0$.
13. R^2 has the exponential distribution with parameter $(2\sigma^2)^{-1}$.
14. Use Theorem 6E with $g(x, y) = \sqrt{(x^2 + y^2)}$ and change to polar coordinates. The variance equals $\sigma^2(2 - \tfrac{1}{2}\pi)$.
15. We see little of virtue in this question. Draw the regions in question in the (x, y)-plane. It is useful to prove that $R = \sqrt{(X^2 + Y^2)}$ and $\theta = \tan^{-1}(Y/X)$ are independent, having the exponential and uniform distributions, respectively.
   (i) $1 - \exp(-\tfrac{1}{2}a^2)$.
   (ii) $\alpha/(2\pi)$.
   (iii) This is rather horrible.
16. (a) (i) $1 - e^{-\lambda x}$ is uniformly distributed on $(0, 1)$.
       (ii) $\min\{X, Y\}$ has the exponential distribution with parameter 2λ.
       (iii) $X - Y$ has the bilateral exponential distribution.
   (b) The answer is 0 if $a < 1$ and $a(1 + a)^{-1}$ if $a \geq 1$.
17. This is largely an exercise in changes of variables, but there is a much better argument which shows $\tfrac{1}{2}$ to be the answer to (ii).

Chapter 7

2. The identity $\sum (X_i - \bar{X})^2 = \sum [X_i - \mu - (\bar{X} - \mu)]^2 = \sum (X_i - \mu)^2 - n(\bar{X} - \mu)^2$ may be useful.

3. $E(S_n/S_n) = nE(X_1/S_n)$. $E(S_m/S_n) = mE(X_1/S_n)$. The result is generally false if $m > n$.

4. $|\{x : F(x) - \lim_{y \uparrow x} F(y) > n^{-1}\}| < n$.

6. Use moment generating functions.

7. For the middle part, find the moment generating function of X_1^2 and use Theorem 7E.

8. This is basically the same argument as in Theorem 4D. $M_S(t) = G_N(M_X(t))$.

9. $\text{var}(Z) = \sum a_n v_{nm} a_m$.

10. Let X_i be 1 if A wins the ith game, and 0 otherwise.

11. $M(s\sigma_1, t\sigma_2)\exp(s\mu_1 + t\mu_2)$.

12. To do the last part, show first that $\psi(t) = M(t)/M(-t)$ satisfies $\psi(t) = \psi(\frac{1}{2}t)^2 = \psi(2^{-n}t)^{2^n}$. Show that $\psi(t) = 1 + o(t^2)$, and deduce that $\psi(t) = 1$ by taking the limit above as $n \to \infty$. Hence $M(t) = M(-t)$, and the original equation becomes $M(t) = M(\frac{1}{2}t)^4$. Repeat the procedure to obtain $M(t) = e^{\frac{1}{2}t^2}$.

13. (ii) Remember the result of Problem 8.
    (iii) XY, where Y has which distribution?

14. Remember Problem 9 in Section 4.5.

15. This is similar to Problem 13.

16. $e^{itx} = \cos tx + i \sin tx$.

17. Use the Inversion Theorem 7J, or remember the density function and characteristic function of the Cauchy distribution.

19. $\phi''(0) = 0$ if $\alpha \geq 2$. What does this imply about such a distribution?

Chapter 8

1. $P(Z_n \leq b) = (b/a)^n$ for $0 < b < a$.

3. Use Theorem 8B.

4. The left-hand side equals $P(X_1 + X_2 + \cdots + X_n \leq n)$ for appropriately distributed random variables.

6. Adapt Example 10 in Section 8.2.

7. If $x + y > \varepsilon$ then either $x > \frac{1}{2}\varepsilon$ or $y > \frac{1}{2}\varepsilon$.

8. Let $Z_i = X_i - Y_i$.

11. $|X| \leq a(1 - I) + MI$ where I is the indicator function of the event that $|X| \geq a$.

12. Combine the results of Problems 10 and 11.

13. Use the Cauchy–Schwarz inequality.

14. Moment generating functions may be useful for the first part. Y has the χ^2 distribution. $E(Y) = 1$, $\text{var}(Y) = 2$.

15. Express $P(Z_n \leq x)$ in terms of the density function of Y_n and take the limit as $n \to \infty$. It is rather complicated.

16. $B_n(p) = \sum_k f(k/n)\binom{n}{k}p^k q^{n-k}$ where $q = 1 - p$. For the last part, note that $n^{-1} \sum X_i$ converges to p, so that, by continuity, $B_n(p)$ is near $f(p)$ with large probability. The error probability is dealt with by using the boundedness of f.

18. The exponential distribution.
19. True, false, and true.
22. This is very similar to Problem 12 in Section 7.6.

Chapter 9

1. S has variance $\sigma^2 \mu^{n-1}(\mu^n - 1)/(\mu - 1)$ if $\mu \neq 1$ and $n\sigma^2$ if $\mu = 1$.
4. $f(s) = e^{\lambda(s-1)}$.

Chapter 10

1. The first part is a variation of the result of Theorem 10A. For the second part, why not follow the first walk for the first n steps and then return along the second walk, reversed, to arrive back at the origin.
2. Either solve the difference equation, or relate it to (15). The number of stages required is $N_1 + N_2 + \cdots + N_M$ where M is the number of moves (remember Theorem 10D) and each N_i has the geometric distribution. Alternatively, solve the appropriate difference equation.
3. This is Problem 2 again. The second difference equation is $t_k = 1 + \frac{1}{6}t_{k-1} + \frac{1}{2}t_k + \frac{1}{3}t_{k+1}$.
4. $1/N$.
5. There is only one solution to $p\theta^3 - \theta + q = 0$ which is less than 1 in absolute value, if $p \leq \frac{1}{3}$.
6. Including any jumps of the walk from N to N we have that if $p \neq q$

$$e(a) = \frac{1}{q-p}\left\{\frac{p}{p-q}[(p/q)^{N-a}(p/q)^N] + a\right\}.$$

7. Use Theorem 10B.
8. For the final part, the general argument in Problem 7 may be useful.
10. Either condition on the value of D_{n-1}, or write $D_n^2 = X_n^2 + Y_n^2$, in the obvious notation.
11. $\{[2s + \sqrt{(4s^2 - s^3 + s)}]^n - [2s - \sqrt{(4s^2 - s^3 + s)}]^n\}[2(s + 1)]^{-n}$. The remaining part of the question is rather similar to Problem 9.
12. This is the three-dimensional version of Problem 8.
15. Use the result of Problem 14.

Chapter 11

1. Poisson, with parameter $\lambda(t - s)$.
2. Condition on N_s.
3. $\frac{1}{6}$.
4. To obtain the integral equation, condition on X_1. The solution of the integral equation is $g(t, x) = e^{-\lambda x}$.
5. Use Theorem 4D.
8. The time between the ith and $(i + 1)$th birth has the exponential distribution with parameter λi.
10. Find $P(T \leq t) = P(L_t = 0)$ from the result of Theorem 11E.
11. To find the mean $m(t)$, differentiate throughout with respect to z and set $z = 1$ to obtain

$$\left[\frac{\partial^2 \phi}{\partial z\, \partial t}\right]_{z=1} = \phi(1, t),$$

giving $m'(t) = 1$ since $m(t) = [\partial\phi/\partial z]_{z=1}$. Hence $m(t) = t$. The variance may be found similarly.

14. The following argument is not completely rigorous but is illuminating. Let $t \to \infty$ in the formula for $G(s, t)$ given in Problem 13 to obtain $G(s, t) \to \exp[\rho(s - 1)]$ where $\rho = \theta/\mu$. This is the probability generating function of the Poisson distribution. Rewrite this in terms of characteristic functions and appeal to the continuity theorem for a watertight argument.

15. $p(t) = e^{-\lambda t} \cosh \lambda t = q(t)$. The time to the first change of state has the exponential distribution with parameter λ. Use independence for the last part.

16. Condition on the length of the service time.

Reading list

Chung, K. L. (1974) *A course in probability theory* (2nd edition). Academic Press, New York.

Cox, D. R., and Miller, H. D. (1965) *The theory of stochastic processes*. Chapman and Hall, London.

Feller, W. (1968) *An introduction to probability theory and its applications,* vol. 1 (3rd edition). John Wiley, New York.

Feller, W. (1971) *An introduction to probability theory and its applications,* vol. 2 (2nd edition). John Wiley, New York.

Grimmett, G. R., and Stirzaker, D. R. (1982) *Probability and random processes*. Clarendon Press, Oxford.

Hall, M. (1967) *Combinatorial theory*. Blaisdell, Waltham, Massachusetts.

Karlin, S., and Taylor, H. M. (1975) *A first course in stochastic processes*. Academic Press, New York.

Index